Edible Insects in the Food Sector

Giovanni Sogari • Cristina Mora
Davide Menozzi

Editors

Edible Insects in the Food Sector

Methods, Current Applications and Perspectives

 Springer

Editors
Giovanni Sogari
Department of Food and Drug
University of Parma
Parma, Italy

Cristina Mora
Department of Food and Drug
University of Parma
Parma, Italy

Davide Menozzi
Department of Food and Drug
University of Parma
Parma, Italy

ISBN 978-3-030-22524-7   ISBN 978-3-030-22522-3 (eBook)
https://doi.org/10.1007/978-3-030-22522-3

This Springer imprint is published by the registered company Springer Nature Switzerland AG
The registered company address is: Gewerbestrasse 11, 6330 Cham, Switzerland

In memory of Marianne Shockley

Preface

The research behind novel food is relatively young and includes a multitude of scientific disciplines and interests (safety, production, nutrition, consumer behavior, sensory analysis, regulations, etc.). New market opportunities for novel products, often derived from nontraditional protein sources, are increasing worldwide; however, consumer acceptability and trust for such alternative foods are often lacking.

This book will explore one of the most discussed and investigated novel foods in recent years: edible insects. Increasing demand for alternative protein sources worldwide had led the Food and Agriculture Organization of the United Nations (FAO) to promote the potential of using insects both for feed and food, establishing a program called "Edible Insects". Although several social, environmental, and nutritional benefits of the use of insects in the human diet have been identified, the majority of the population in Western countries rejects the idea of adopting insects as food, predominantly for cultural reasons.

Nevertheless, since the 2013 publication of the FAO report entitled, "Edible Insects: Future Prospects for Food and Feed Security," international interest in promoting the consumption of insects has grown significantly, mainly in North America and Europe. This trend is mostly due to increasing attention and involvement from research institutes, the food and feed industries, as well as governments and their constituents. For instance, in recent years, an increasing number of start-ups were born in the food industry aiming to develop insect-based products, while large companies have sought to invest in this sector.

This book covers the current state of entomophagy, taking into consideration the consumer point of view, aspects of safety and allergies for human consumption, final meat quality of animals fed with insects, the legislative framework for the commercialization of this novel food, and other relevant issues.

One of the aims is to identify knowledge gaps to inform primary research institutions and other stakeholders, like funding organizations, to better plan, develop, and implement future research activities on edible insects as a sustainable source of food and feed.

The book is composed of eight themed chapters. The first opens with an outlook of the risk assessment and the future prospects of insects as food in Europe. Tilemachos Goumperis presents how the European Food Safety Authority (EFSA), as well as national authorities, has published assessments and guidelines discussing farming, processing, and consumption of this "novel food."

In the second chapter, Charlotte Payne, Rudy Caparros Megido, Darja Dobermann, Francis Frédéric, Marianne Shockley, and Giovanni Sogari address the question of why the Global North, Europe and North America, has given a novel and neglected food group so much attention lately and how multiple sectors—academia, media, and industry—have begun to popularize and call for a resurgence in the use of insects as food.

In the following chapter, Giovanni Sogari, Davide Menozzi, Christina Hartmann, and Cristina Mora highlight the overall state of primary research activities and trends related to eating insects and the associated consumer behavior, describing and summarizing the characteristics and methodological approaches arising from different quantitative and qualitative studies worldwide.

The fourth chapter deals with the drivers and barriers of consumer acceptance of insects as food. Hartmann and Bearth explore the role of emotional reactions toward edible insects, such as disgust, and motivational barriers to acceptance, as well as other factors like risk, benefit perception, and trust.

In the fifth chapter, Luís Miguel Cunha and José Carlos Ribeiro provide a review of the main factors underlying entomophagy rejection and acceptance, with a focus on sensory properties appeal of insects and insect-containing products. Strategies to increase Western consumers' acceptance have been addressed.

The sixth chapter, authored by Laura Gasco, Ilaria Biasato, Sihem Dabbou, Achille Schiavone, and Francesco Gai, analyzes some aspects of quality and con-sumer acceptance of products from animals fed insect meals, critically reviewing the latest knowledge about the dietary use of insect meals in fish, shellfish, and avian species, also in terms of sensorial perception.

In chapter seven, José Carlos Ribeiro, Luís Miguel Cunha, Bernardo Sousa-Pinto, and João Fonseca explore the potential allergenic risks of entomophagy, reviewing the molecular mechanisms implied in cross-reactivity/co-sensitization and describing case studies on allergic reactions following the intentional consump-tion of insects.

Finally, the last chapter gives a complete overview of recent legal frameworks for insects as food. Francesca Lotta presents the main elements and differences of the regulatory classification of insects as food in Europe (the Novel Food Regulation) and the United States, as well as the rules food operators need to comply with to legally place these products on the market.

We hope this book will support stakeholders, including scientific networks, stu-dents, members of the private sector, and policymakers, who are interested in learn-ing more about this novel food and increasing their expertise and know-how on the methods and approaches to study entomophagy.

The editors wish to thank the authors of this volume—all renowned scientists and experts in the field of edible insects—for their valuable contributions; the publication of this book would not have been possible without their efforts, hard work, and expertise.

Parma, Italy

Giovanni Sogari
Cristina Mora
Davide Menozzi

Contents

About the Editors

Giovanni Sogari, PhD, Giovanni is a Postdoctoral Researcher in the Department of Food and Drug at the University of Parma, Italy, and currently a visiting scholar in the Charles H. Dyson School of Applied Economics and Management, at Cornell University in Ithaca, NY, USA. He was a recipient of a Marie Skłodowska-Curie Fellowship in 2017 (EU Horizon 2020 program) and he is currently working on his project "CONSUMEHealth: Using consumer science to improve healthy eating habits." The objective is to understand what drives consumers to make healthier food choices and provide evidence-based recommendations for stakeholders and policymakers. His other main research interests are consumer behavior about geographical indications (GI), food produced from genetically modified organisms (GMOs), the wine market, sustainability in the food system, and novel foods, including edible insects. He is a member in various scientific societies including the Italian Scientists and Scholars in North America Foundation, the Institute of Food Technologists (IFT), the Agricultural and Applied Economics Association (AAEA), the Italian Association of Agricultural and Applied Economists (AIEAA), and the European Association of Agricultural Economists (EAAE). You can follow him at: www.giovannisogari.com

Cristina Mora, PhD, Cristina is a Full Professor in the Department of Food and Drug at the University of Parma, Italy. She has done extensive research to understand individual responses to risks and benefits associated with food, health, sustainability, and safety. Additionally, her work includes a growing body of literature on qualitative methodologies to understand attitudes and behaviors related to emerging food production technologies and sustainability, as well as stakeholder analysis and public engagement in the agrifood sector. She has been involved in several national and EU research projects, primarily in the area of social science and the (agri)food sector, and has been work package (WP) leader in many of these. Recent examples include Focus Balkans, Pegasus, and Prime Fish projects (www.primefish.eu). Cristina also undertakes teaching activities at the undergraduate and postgraduate level in Italy on consumer behavior and agribusiness. She is a member of the Italian Association of Agricultural and Applied Economists (AIEAA) and the European Association of Agricultural Economists (EAAE).

Davide Menozzi, PhD, Davide is an Associate Professor at the University of Parma, Italy, in Agricultural Economics and Rural Appraisal. He has been involved in several national and EU research projects since 2000, focusing on the analysis of consumer behavior and preferences, the economic analysis of food safety and food quality schemes with a supply chain perspective, and the evaluation of the socioeconomic sustainability of dietary behaviors. He teaches graduate and undergraduate courses, including "Agricultural and Food Economics" and "Food Choice and Consumer Behavior," in the Department of Food and Drug and the Department of Economics and Management at the University of Parma. He obtained the National Scientific Qualification to the functions of Full Professor in the field of Agricultural Economics and Rural Appraisal in March 2018. He is an active member in various scientific societies including the Italian Association of Agricultural and Applied Economists (AIEAA) and the European Association of Agricultural Economists (EAAE).

Chapter 1
Insects as Food: Risk Assessment and Their Future Perspective in Europe

Tilemachos Goumperis

Abstract Whilst insects as food is still a niche market in Europe, the interest of consumers and industry has increased in the last years as insects are seen as an alternative source of protein with nutritional and economic benefits. At the same time, farming, processing and consumption of insects and products thereof as food may pose risks and, therefore, the European Food Safety Authority (EFSA), as well as, national authorities published assessments or guidelines discussing these aspects. According to the EU regulatory framework, insects and their products are considered "novel foods". They can be marketed only if authorised, a process that implies a safety assessment. EFSA guidance documents detail the data needed for the safety assessment.

Keywords Insects · Food · Novel · Risk assessment

Introduction

There is an increasing interest to farm and use insects as food during the last years. Insects have been suggested as potent bio-converters, which can transform low quality and cheap biomass into nutritionally valuable and economically profitable proteins. The term "entomophagy", from Greek "entomo" meaning insect and "phagia" meaning to eat, has been proposed to refer to the consumption of insects as food. Eating insects may be unpleasant to some people and the "yuck factor" among consumers may be the major reason for rejecting insects as food. In western societies insects are only related to bad hygiene and biological contamination, but in contrast, insects are part of the staple diet in some countries in Asia, Africa and

Disclaimer: The views expressed in this publication are those from the author and do not necessarily represent the official position of EFSA. EFSA assumes no responsibility or liability for any errors or inaccuracies that may appear.

T. Goumperis (✉)
European Food Safety Authority (EFSA), Parma, Italy
e-mail: tilemachos.goumperis@efsa.europa.eu

South America. Discrepancies in the way people like or dislike the idea of consuming insects can be found also within the same country; for example in Northern Thailand, insects are produced in insect farms and regularly consumed, whereas in the capital, Bangkok, insects are consumed in lower quantities.

Publication of articles on scientific journals and media on edible insects is on the rise, discussing among other issues nutritional, societal, environmental, safety and production aspects. Scientific conferences and scientific symposia dedicated to edible insects are now regularly organised (such as the conferences "Insects to feed the world" held in 2014 and 2018, the annually organised INSECTA conferences and the Symposium of the German Federal Institute for Risk Assessment (BfR) (Schafer et al. 2016).

Economic Factors

Currently, insects as food is a niche market in the EU. Insect products already on the market include whole insects, such as dried mealworms with mint flavour and grounded crickets (cricket flour) as an ingredient in pasta.

Start-up companies established during the last years, produce insects or products deriving from insects as food, as well as, feed. Industry clusters have been founded to promote production and research on insects. These include the Belgian Insect Industry Federation (BIIF; http://www.biif.org/) and the International Platform of Insects for Food and Feed (IPIFF; http://ipiff.org/) based in the EU and the Asian Food and Feed Insect Association (AFFIA; https://affia.org/).

Small-scale insect farming may also impact the local economy. An example of this is Thailand, where insect farming is well developed with 20,000 insect farming enterprises registered in 2013 most of which are household small-scale farms (FAO 2013a) producing two types of edible insects, cricket and palm weevil larvae in the north and south on the country respectively. Another example comes from Laos, where small-scale farming is a tool to increase family income, as well as, to improve household nutrition in Laos (Weigel et al. 2018).

Nutritional Aspects

Whilst it was stated that more than 1900 insect species have been documented in the literature as edible (FAO 2013b; Jongema 2017), in practice, approximately only dozen of them are in the radar of insect farmers and risk assessors (EFSA 2015).

The nutritional profile of insects is highly variable depending on the species themselves, their developmental stage in time of harvesting (that can span from eggs to larvae, pupae or adults) and the substrate (insects' feed) that is used to rear the insects on. EFSA (2015) reported a comprehensive summary of data on nutritional aspects. Antinutritional substances have been identified in some insect species too (ANSES 2015) such as phytic acid and tannins.

Insects have been suggested as valuable source of protein; for example, the protein content of mealworm (*Tenebrio molitor*) has been reported to count from 47% to 60% in dry matter (Makkar et al. 2014). Because of their high content of protein/ amino acids, fibre and micronutrients, they could be an alternative to traditional food of animal origin.

Handling and processing of insects may also impact their nutritional composition. For example, rapid oxidation of insect products may happen during processing due to their high unsaturated fatty acid content (FAO 2013b).

EU National Risk Assessments and Guidelines

As any other food, consumption of insects and of products thereof could pose food safety risks. In Europe, national food safety authorities performed risk assessments for insects. These assessments discussed potential microbiological and chemical risks, allergenicity and the effect of processing (AECOSAN 2018; ANSES 2015; FASFC 2014; NVWA 2014).

The Finnish Food Safety Authority (Evira 2018) and the Austrian authorities (Austrian Ministry for Health 2017) prepared guidelines covering aspects of farming, marketing and food safety issues of insects as food. Adequate hygiene measures and a system based on Hazard Analysis and Critical Control Points (HACCP) have been proposed as prerequisites during breeding, processing and marketing of insects for human consumption.

Risk Assessments Outside EU

In Switzerland, whole animals or flour of three species can be marketed: *Acheta domesticus* in adult phase; *Locusta migratoria* in adult phase; and larvae of *T. molitor* (FOAG 2017). However, protein extracts arriving from these insects have not been permitted.

The Food Standards Australia and New Zealand Advisory Committee on Novel Foods (ACNF) undertook an assessment of three insect species for human consumption (*Zophobas morio* (super mealworm), *A. domesticus* (house crickets), *T. molitor* (mealworm)) concluding that there are no safety concerns for human consumption (FSANZ 2018).

In Thailand, good agricultural farming practices for cricket farming have been published (ACFS 2017).

EFSA Opinion on a Risk Profile of Insects as Food and Feed

The European Food Safety Authority assessed the risks related to the consumption of insects as food, as well as feed (EFSA Scientific Committee 2015). Its Opinion discussed microbiological, chemical and environmental hazards, as well as allergenicity, arising from primary production (insect farming) processing and consumption. The substrate used during farming of insects was associated with the probability of the occurrence of hazards in insects and products thereof.

In more detail, with regards to microbiological hazards, pathogenic bacteria (such as *Salmonella* and *Campylobacter*) and viruses may be present in non-processed insects, but the risk of transmission to the final product can be mitigated by the application of effective processing. With regards to prion-related risks, the Opinion concludes that, whilst no insect-specific prion diseases can develop in insects, they could act as mechanical vectors of infection. Accumulation of heavy metals in insects from the substrate was also shown, in particular cadmium.

Allergic reactions, including anaphylactic shocks in humans, have been reported. It was recommended that labelling can be used as a mitigation measure to alert the consumers for the presence of insect protein and the possible allergenicity or cross reactivity to allergens such as tropomyosin or arginine kinase.

The Opinion also makes a comparison of hazards in non-processed insects with the occurrence of these hazards in other protein sources of animal origin. This comparison is made on the basis of rearing insects in seven different substrate groups, such as, authorised feed materials according to the EU catalogue of feed materials (Regulation (EU) No 68/2013), food produced for human consumption, but which is no longer intended for human consumption (e.g. due to expired use-by date), catering waste or animal manure.

The EFSA Opinion points out the need for risk assessments on different species and substrate taking into account the whole production chain from farming to processing and to consumption. A number of areas of uncertainty were highlighted, including the environmental impact of mass-rearing production and precise details of the production processes used for different insects and by different producers.

Insect Species Specific Assessments

Food safety risks may vary considerably among insect species. Whilst the above-mentioned generic assessments can be used as guidelines, more targeted risk assessments need to be performed at the level of insect species. Fernandez-Cassi et al. (2018) published a risk profile for the house cricket (*A. domesticus*). The authors screened existing literature specifically for this species and identified potential hazards. In case of data gaps, evidence available for other species of the Orthoptera genus was used (e.g. grasshoppers, locusts and other cricket species). The identified risks were ranked in a scale of low, medium and high considering the probability of the existence of the hazard and the consequences from the exposure. The results of this exercise categorised the following risks as medium and high: *(i) high total aerobic bacterial counts; (ii) survival of spore-forming bacteria following thermal processing; (iii) allergenicity of insects and insect-derived products; and (iv) the bioaccumulation of heavy metals (e.g. cadmium)"*.

Novel Foods Regulation and Relevant EFSA Guidance Documents

Regulation (EC) No 258/97 on novel foods has been into force until the end of 2017. In relation to the definition of the novel foods, Article 1(2)(e) as regards to animals referred to *"food ingredients isolated from animals"*, whereas as regards to plants to *"food ingredients consisting of or isolated from plants"*. Food ingredients extracted or isolated from insects (e.g. protein isolates) fell within the definition of novel food as food ingredients isolated from animals. Similarly, insects for which parts have been removed (such as legs, wings, head, intestines etc.) fell within the same definition. However, in some EU Member States, it was debated on whether whole insects (e.g. grasshoppers) being whole animals and preparations made of whole insects (e.g. grasshopper flour) were also included. The result of this debate was that in some EU Member States whole insects and their preparations were allowed to be marketed as there were not considered novel foods.

Regulation (EU) 2015/2283 on novel foods came into force on 1 January 2018. The definition of novel foods in the EU is given in Article 3(a):

> *"'novel food' means any food that was not used for human consumption to a significant degree within the Union before 15 May 1997, irrespective of the dates of accession of Member States to the Union, […]"*;

where point (v) reads: *"food consisting of, isolated from or produced from animals or their parts, except for animals obtained by traditional breeding practices which have been used for food production within the Union before 15 May 1997 and the food from those animals has a history of safe food use within the Union"*.

With the application of this Regulation, whole insects and their products are considered as novel foods. Therefore, food business operators can place such products on the EU market only if they are authorised. The administrative and scientific requirements for novel food applications are further detailed in the Commission implementing Regulation (EU) 2017/2469.

As for any other novel food, if insects or insect products become authorised novel foods, they will be included in the "Union list of novel foods". The Union list is updated by the European Commission and includes their conditions of use, labelling requirements and their specifications.

In 2018, the European Commission received a number of applications related to insects (EC 2018). These included *A. domesticus* (house cricket); whole and ground *Alphitobius diaperinus* (lesser mealworm) larvae products; dried *Gryllodes sigillatus* (cricket); migratory locust (*L. migratoria*); dried *T. molitor* (mealworm).

Regulation (EU) 2015/2283 also foresees that the European Commission shall request the opinion of EFSA to assess the safety of novel foods. For this purpose, EFSA published a guidance document to assist the applicants in preparing an application for authorisation of a novel food in the context of Regulation (EU) 2015/2283 to demonstrate its safety (EFSA NDA Panel 2016a). Data needed for the safety assessment include the description of the novel food and its production process, compositional data, proposed uses and use levels. In addition, the history of use of the novel food and/or its source, nutritional information, toxicokinetics (absorption, distribution, metabolism and excretion), toxicodynamics and allergenicity should be considered by the applicant by default or if not, this should be justified. On the basis of the information provided by the applicant, EFSA assesses the safety of the novel food under the proposed conditions of use and anticipated intake. According to Regulation (EU) 2015/2283, EFSA shall provide its opinion within a period of nine months that can be extended in cases where EFSA requests additional information from the applicant.

Insects and their products may be considered as traditional food in a third (non-EU) country. Regulation (EU) 2015/2283 covers also this case and lays down the process for authorisation of a traditional food from a third country to be placed in the EU market. In this situation, an applicant must submit to the European Commission a notification demonstrating, among other information, the history of safe food use in the third country. EFSA published a guidance document on the preparation and presentation of the notification of traditional foods from third countries in the context of the novel food Regulation (EFSA NDA Panel 2016b). Data needed for the safety assessment include information on the description, production process, composition, stability data and specifications of the traditional food and the proposed conditions of use. The traditional food will be authorised to be marketed, unless, EU Member States or EFSA submit duly reasoned safety objections within a period of four months.

Future of Entomophagy in Europe

Would the EU population introduce insects into their diet? To answer this question, (at least) three factors have to be discussed.

The first factor is compliance with the legal framework pertaining the food production and supply. A number of applications for authorisation of different insect species as novel foods or traditional foods from countries outside the EU have already been submitted to the relevant EU authorities and more may follow in the future. Once the safety assessments are concluded and if it is shown that their production and consumption is safe, some of these insect species, if not all, might soon be authorised to be sold as whole foods or food ingredients in the EU. In case the scientific evidence to perform the risk assessment is not sufficient for an insect species, demonstrating additional evidence may be needed before the final decision is made.

Apart from authorisation matters, farming of insects and production of foodstuff based on insects have also to respect the relevant legislation, including (but not limited to) the following areas:

- The feedstuff materials that can be used for insect farming. In particular, the current feed EU legislation includes a list of "prohibited material", including household waste and faeces, that cannot be used for farming animals (Regulation (EC) No 767/2009; Article 6 and Annex III).
- Hygiene during farming and processing as for any other foodstuff intended for human consumption.
- Maximum limits for undesirable substances such as contaminants (e.g. mycotoxins or heavy metals), residues of veterinary medicines and plant protection products.
- Microbiological criteria for foodstuffs.
- Importation from countries outside the EU.

Another issue is if animal welfare should be considered for insects during production and killing. Scientific evidence to whether insects feel pain is not conclusive or at least is lacking for the time being.

The second factor lies on having an economically viable production that can provide enough volume of insects all year round. Several companies have started producing insects for use as food. These companies where either producing insects as pet food before and extended their business to a new marketing opportunity or were recently established viewing a potential in the food market. The time will show how profitable insect farming and/or manufacturing of insect products can be.

The third factor is the selection of the right products to be marketed and their acceptance by the consumers. Food culture in Europe has long history and is linked to social traditions and stereotypes. Whilst one may agree that the use of insects brings important environmental, economic and food security benefits, as well as that insects are nutrient dense, they may disagree that such a "revolution" to introduce insects in the diet of Europeans should commence from themselves. Market research and new product development projects of the insect production industry shall be able to demonstrate which insect species and which insect-based products can offer the advantages they claim, in a competitive price and in a way that can be accepted by the European consumers.

References

ACFS (Thailand's National Bureau of Agricultural Commodity and Food Standards) (2017) Good agricultural farming practices, guidance on the application of Thai agricultural standard TAS 8202(g)-2017. http://www.acfs.go.th/standard/download/GUIDANCE_GAP_CRICKET_FARM.pdf. Accessed 17 Sept 2018

AECOSAN (2018) Report of the Scientific Committee of the Spanish Agency for Consumer Affairs, Food Safety and Nutrition (AECOSAN) on the microbiological risks associated to insect consumption. http://www.aecosan.msssi.gob.es/AECOSAN/docs/documentos/

seguridad_alimentaria/evaluacion_riesgos/informes_comite/CONSUMO_INSECTOS.pdf. Accessed 17 Sept 2018

ANSES (French Agency for Food, Environmental and Occupational Health and Safety) (2015) Opinion on the use of insects as food and feed and the review of scientific knowledge on the health risks related to the consumption of insects. https://www.anses.fr/en/documents/BIORISK2014sa0153EN.pdf. Accessed 17 Sept 2018

Austrian Ministry for Health (2017) Guidelines for farmed insects as food. https://www.verbrauchergesundheit.gv.at/tiere/publikationen/Futterinsekten.html. Accessed 17 Sept 2018

EFSA NDA Panel (EFSA Panel on Dietetic Products, Nutrition and Allergies), Turck D, Bresson J-L, Burlingame B, Dean T, Fairweather-Tait S, Heinonen M, Hirsch-Ernst KI, Mangelsdorf I, McArdle H, Naska A, Neuhäuser-Berthold M, Nowicka G, Pentieva K, Sanz Y, Siani A, Sjödin A, Stern M, Tome D, Vinceti M, Willatts P, Engel K-H, Marchelli R, Pöting A, Poulsen M, Salminen S, Schlatter J, Arcella D, Gelbmann W, de Sesmaisons-Lecarre A, Verhagen H, van Loveren H (2016a) Guidance on the preparation and presentation of an application for authorisation of a novel food in the context of Regulation (EU) 2015/2283. EFSA J 14(11):4594. https://doi.org/10.2903/j.efsa.2016.4594

EFSA NDA Panel (EFSA Panel on Dietetic Products, Nutrition and Allergies), Turck D, Bresson J-L, Burlingame B, Dean T, Fairweather-Tait S, Heinonen M, Hirsch-Ernst KI, Mangelsdorf I, McArdle H, Naska A, Neuhäuser-Berthold M, Nowicka G, Pentieva K, Sanz Y, Siani A, Sjödin A, Stern M, Tome D, Vinceti M, Willatts P, Engel K-H, Marchelli R, Pöting A, Poulsen M, Schlatter J, Gelbmann W, de Sesmaisons-Lecarre A, Verhagen H, van Loveren H (2016b) Guidance on the preparation and presentation of the notification and application for authorisation of traditional foods from third countries in the context of Regulation (EU) 2015/2283. EFSA J 14(11):4590. https://doi.org/10.2903/j.efsa.2016.4590

EFSA Scientific Committee (2015) Scientific Opinion on a risk profile related to production and consumption of insects as food and feed. EFSA J 13(10):4257., 60 pp. https://doi.org/10.2903/j.efsa.2015.4257

European Commission (2018) Summary of ongoing applications and notifications: https://ec.europa.eu/food/safety/novel_food/authorisations/summary-ongoing-applications-and-notifications_en. Accessed 17 Sept 2018

Evira (Finnish Food Safety Authority) (2018) Guide Insects as food (10588/2). https://www.evira.fi/globalassets/tietoa-evirasta/lomakkeet-ja-ohjeet2/elintarvikkeet/eviran_ohje_10588_2_uk.pdf. Accessed 17 Sept 2018

FAO (Food and Agriculture Organization of the United Nations) (2013a) Six-legged Livestock: Edible insect farming, collection and marketing in Thailand. ISBN 978-92-5-107578-4

FAO (Food and Agriculture Organization of the United Nations) (2013b) Edible insects. Future prospects for food and feed security. van Huis A, van Itterbeeck J, Klunder H, Mertens E, Halloran A, Muir G and Vantomme P. http://www.fao.org/docrep/018/i3253e/i3253e00.htm. Accessed 17 Sept 2018

FASFC (Belgian Scientific Committee of the Federal Agency for the Safety of the Food Chain) (2014) Food safety aspects of insects intended for human consumption. Common advice of the Belgian Scientific Committee of the Federal Agency for the Safety of the Food Chain (FASFC) and of the Superior Health Council (SHC). http://www.favv-afsca.fgov.be/scientificcommittee/advices/_documents/ADVICE14-2014_ENG_DOSSIER2014-04.pdf. Accessed 17 Sept 2018

Fernandez-Cassi X, Supenu A, Jansson A, Boqvist S, Vagsholm I (2018) Swedish University of Agricultural Sciences (SLU), Department of Biomedical Sciences and Veterinary Public Health, Sweden. Novel foods: a risk profile for the house cricket (*Acheta domesticus*). EFSA J 16(S1):e16082. 15 pp.

FOAG (Federal Office for Agriculture) (2017) Information letter 2017/1; production and processing of insects for use as foodstuffs. https://www.blw.admin.ch/blw/it/home/nachhaltige-produktion/tierische-produktion/insekten.html. Accessed 17 Sept 2018

FSANZ (Food Standards Australia New Zealand) (2018) Record of views formed in response to inquiries (updated October 2018). http://www.foodstandards.gov.au/industry/novel/novelrecs/SiteAssets/Pages/default/Record%20of%20views.pdf. Accessed 15 Nov 2018

Jongema Y (2017) List of edible insects of the world. Laboratory of Entomology, Wageningen University, Wageningen, the Netherlands. https://www.wur.nl/en/Research-Results/Chair-groups/Plant-Sciences/Laboratory-of-Entomology/Edible-insects/Worldwide-species-list.htm. Accessed 1 Sept 2018

Makkar HPS, Tran G, Heuze V, Ankers P (2014) State-of-the art on use of insects in animal feed. Animal Feed Sc Tech 197:1–33

NVWA (Netherlands Food and Consumer Product Safety Authority) (2014) Advisory report on the risks associated with the consumption of mass-reared insects. http://www.nvwa.nl/actueel/risicobeoordelingen/bestand/2207475/consumptie-gekweekte-insecten-advies-buro. Accessed 17 Sept 2018

Schafer B, Bandick N, Epp A, Hirsch-Ernst KI, Pucher J, Schumann R, Spolders M, Wagner B, Lampen A (2016) BfR Symposium 'Insekten als Lebens- oder Futtermittel: Nahrung der Zukunft?' Bericht zum Symposium am 24 Mai 2016. J Verbr Lebensm 11:281. https://link.springer.com/article/10.1007%2Fs00003-016-1038-0. Accessed 17 Sept 2018

Weigel T, Fèvre S, Berti PR, Sychareun V, Thammavongsa V, Dobson E, Kongmanila D (2018) The impact of small-scale cricket farming on household nutrition in Laos. J Insects Food Feed 4(2):89–99

Chapter 2
Insects as Food in the Global North – The Evolution of the Entomophagy Movement

Charlotte Payne, Rudy Caparros Megido, Darja Dobermann, Francis Frédéric, Marianne Shockley, and Giovanni Sogari

Abstract The last decade has seen a surge of interest and investment in insects as food and feed. Has the Global North ever before given a novel and neglected food group so much attention? In this chapter we describe how and why multiple sectors - academia, media, industry - have begun to popularise and call for a resurgence in the use of insects as food in Europe and the US, despite tenacious taboos.

We begin with an overview of the history of insect consumption in the Global North; indigenous peoples in regions of both the US and Europe have traditionally consumed insects in some form, but this has diminished, disappeared or even been actively suppressed in recent history.

We describe the beginnings of an active interest in rediscovering insect consumption, beginning with a handful of entomologists who saw the potential of insects as an alternative to meat for reasons of taste, nutrition and environmental impact.

These ideas truly reached the mainstream in 2013 when the FAO published a paper on edible insects. This was immediately picked up by the world's media, by scientists, and by multiple entrepreneurs.

C. Payne (✉)
Department of Zoology, University of Cambridge, Cambridge, UK
e-mail: clrp2@cam.ac.uk

R. Caparros Megido · F. Frédéric
Functional and Evolutionary Entomology, Gembloux Agro-Bio Tech – University of Liège, Gembloux, Belgium

D. Dobermann
Rothamsted Research, Harpenden, UK

M. Shockley
UGA Department of Entomology, GA University of Georgia, Athens, GA, USA

G. Sogari
Department of Food and Drug, University of Parma, Parma, Italy

Charles H. Dyson School of Applied Economics and Management, Cornell University, Ithaca, NY, USA

© Springer Nature Switzerland AG 2019   11
G. Sogari et al. (eds.), *Edible Insects in the Food Sector*,
https://doi.org/10.1007/978-3-030-22522-3_2

Since then, the entomophagy movement has gathered pace. Insects have been hailed as a 'superfood', are widely available to buy online and are increasingly found on the shelves of some retail outlets.

In this chapter we recount this recent historical trajectory. In doing so we also discuss the shifts in societal attitudes towards insects as food, critical gaps in research, and market opportunities for current and future entrepreneurs in the field.

Keywords Edible insects · Entomophagy · Europe · US · Food history · Entrepreneurship

Introduction

Insects are not a 'new' food. We have eaten insects throughout human history (Meyer-Rochow 2010; Evans et al. 2015) and prehistory (Backwell and D'Errico 2001; Pager 1973, 1976), and continue to do so today. High-end cake manufacturers in urban metropolises, subsistence farmers in remote rural areas, and, most recently, social entrepreneurs in suburbia, have this in common: a number of them use insects as a source of nutrition and income. Many red foods are dyed with carmine, which is extracted from the cochineal beetle (Krahl et al. 2016). Carmine is used globally and cochineal beetles are farmed on a large scale in South America, particularly in Peru (Campana et al. 2015). Many subsistence farming communities in warmer climates collect insects from their fields (Payne and Van Itterbeeck 2017) and from surrounding uncultivated land (Yen 2015; Van Huis 2003), to sell and to eat (Manditsera et al. 2018; Kelemu et al. 2015). In several parts of the world, social entrepreneurs have founded businesses that sell insect foods (Fleming 2016; Dunkel and Payne 2016; Müller et al. 2016).

However, this is not how insect-eating began. Evidence suggests that long before agriculture, hominin hunter-gatherers used tools to forage for termites (Lesnik 2018). With the development of agriculture, large herbivorous insects such as crickets and grasshoppers likely proliferated: crops provided nutrient-rich plant food in abundance.

As agriculture spread to increasingly northern climates, only smaller-bodied crop pests persisted. This may be one reason for a relative lack of insect-eating culture in cooler latitudes. In recent history, eating insects has been actively discouraged and tabooed in Europe and the US, and in many areas of the world where these countries have exerted their influence - marginalisations of insect food customs worldwide are among countless examples of colonial actors eroding existing cultural diversity (Yen 2009).

In this chapter we discuss the trajectory of insect consumption in Europe and the US from its roots in traditional cultures to its resurgence in restaurants, in research, and on supermarket shelves.

An Historical Overview

Cicadas, Silkworm and Casu Marzu: Traditional Insect Foods in Europe

The earliest accounts of insects eaten in Europe come from the Greeks and Romans, and by all accounts they were highly valued foods: Greek historians report that cicadas were eaten at banquets (Bodenheimer 1951, p. 39), and during the time of the Roman empire Capricorn beetle larvae (*Cerambyx scopolli*, prev *Cerambyx heros*, see Bodenheimer 1951, p. 43) were fed with flour and wine to fatten them before they were eaten (Evans et al. 2015).

Although no records suggest that insects were a common food in Europe following the decline of the Roman Empire, there are scattered accounts. Silkworm were farmed for their silk in rural areas of Italy, Germany, Spain and France (Di Vittorio 2006), and in the sixteenth century, the Italian naturalist Androvaldi described German soldiers in Italy who enjoyed eating fried silkworms. Occasional outbreaks of insects may also have prompted consumption: in the late seventeenth century locusts were eaten at a feast after a plague of the insects spread across Germany (Bodenheimer 1951, p.44), and similarly, an outbreak of *Plusia gamma* in late eighteenth century France prompted the French entomologist de Reamur to consider their edibility (DeFoliart 2002). 'Cockchafers' are another large-bodied insect prone to outbreaks and are found in European recipe books from the 19th and 20th centuries (Hyman 2013; Mlček et al. 2018; Ghosh et al. 2018). While we lack evidence showing how commonly this dish was eaten, we do know that prior to the increase in pesticide use in the 1970s, these beetles had very healthy populations across the continent (Warner 2006).

Another insect food tradition that was once found in several European countries is the processing of milk into a cheese known as Casu Marzu (Sardinia), U Casgiumerzu (Corsica), Pecorino Marcetto (Abruzzo, Italy), Trulo Sir (Croatia) or specific types of Queso de Cabrales (Asturias, Spain), using fly larvae (Evans 2018). The larvae produce enzymes that alter the fermentation and flavour of the cheese, and are commonly eaten, live, with the cheese (Manunza 2018). In the later half of the twentieth century, mass urban-rural migration fostered an antipathy to traditional products associated with rural poverty and being 'backward' (Parasecoli 2014); in the case of Casu Marzu, this growing hostility towards it became law when a regulation was passed in 1962 banning its sale (Manunza 2018).

Overall, insect food traditions in Europe do exist, but their survival and acceptance has been sparse, sporadic and in recent decades, dismissed.

Great Basin Grasshoppers and Mormon Crickets: Traditional Insect Foods in the US

Writing in 2002, Defoliart (2002) records 84 species of insect consumed in "North America north of Mexico". This list is almost certainly an underestimate, since for many records the genus name or species name is unrecorded. Historically, we can certainly say that the use of insects as food predates the present of European migrants; archaeological evidence suggests that traditions of insect consumption in North America were present even in the late Pleistocene (Goebel et al. 2011) and early Holocene (Sutton 1995, p. 268). Insect consumption does seem to have been most prevalent in the western and southern regions of the continent, and several species were highly prized as food even in regions with an abundance of wild animals, fish and fruit (DeFoliart 2002).

In North America as in Europe, some outbreak insects were consumed, and some scarcer species were prized for their flavour. However, also in parallel to European trends, both environmental degradation and the erosion of traditional cultures have rendered indigenous insect consumption extremely rare.

In a landmark example of the former, Rocky Mountain grasshoppers were recorded by several historians of the nineteenth century as being a traditional indigenous food used from California up to the Northern states (Defoliart 2002). However, the population crashed in the late nineteenth century, never recovered, and is now extinct (Chapco and Litzenberger 2004). This has been attributed to several aspects of anthropogenic change, including tillage, irrigation, pesticide use, and a changing species assembly due to human disturbance (Lockwood and DeBrey 1990).

The archaeologist Masden – upon finding extensive evidence of the consumption of orthoptera (grasshopper and related species) – evaluated the energetic returns gained from gathering grasshopper and related species in the Western US, and suggested that this could be a more profitable activity than hunting large game since, based on an experimental reconstruction, collectors could expect to gather up to 273,000Kcal per hour (Madsen 1989). Notably, orthopteran insects were also threats to crop-based agriculture: the Mormon cricket, another North American orthopteran once widely enjoyed as food, is so-called because of the threat it posed to the crops of Mormon migrants in 1848 (Madsen 1989).

However, as in Europe, reasons for eating insects were not purely economic nor rational. Many wasp species, including lethal yellowjackets, were also collected, cooked and eaten as delicacies particularly in western and southern regions (DeFoliart 2002). The taste, we assume, was worth the risk.

Overall, traditional insect consumption in North America seems to have been more widespread than in Europe, with a greater number of species consumed. This may be due to the relatively high diversity of cultural traditions and large-bodied insect species on the continent prior to mass migration from Europe.

The Beginnings of the Entomophagy Movement

Entomologists Eating Insects: Bug Banquets and the Food Insects Newsletter

Based on the examples above, we can posit that the 'entomophagy movement' of recent years may be best framed as the revival of a marginalised tradition. This revival is surely rooted in the small but robust contingent of entomologists in both Europe and the US, who began to consider the use of insects as food as a way to mitigate problems caused by human population growth. In 1975, the American entomologist Gene DeFoliart published a paper discussing insects as a source of protein and suggesting research was needed to explore their potential (DeFoliart 1975); in 1976, the Finnish biologist Meyer-Rochow published a short opinion piece that concluded eating insects could help in 'easing the problem of world protein and food shortage' (Meyer-Rochow 1976). Neither was well-cited in the years immediately following their publication, but during the 1980s several research papers examined the nutritional value and environmental impact of certain insects as food. To unite the community of emerging researchers interested in this topic, Gene DeFoliart founded the Food Insects Newsletter in 1988. He did this in collaboration with researchers in tropical countries that had a history of using insects as food after finding that

> *'edible insects are indeed still widely used as food throughout the rural tropical world. In fact, the prevailing opinion among those most knowledgeable about the situation in specific regions is that edible insects not only continue to be nutritionally important but could make an even greater contribution to human nutrition if supplies were increased or better distributed seasonally.'* (DeFoliart 1988)

The rise in interest in the topic demonstrated by the newsletter was also evidenced in the establishment of annual Bug Banquets at Montana State University (MSU) in 1989. During the thirty years in between then and now, interest in the potential of insects as food among the research community in North America grew slowly but steadily: Bug Banquets are still a regular feature in the MSU calendar and now attract over 850 guests (Besemer 2018).

Beyond Taste: Emerging Data on Insect Efficiency and Nutrition

The growing revival of this marginalised tradition has been coupled with a steep increase in research efforts to substantiate claims surrounding production efficacy and nutritional quality of insects; two claims which form the base argument for the role of insects in mitigating food security challenges (Van Huis et al. 2013).

With regard to nutrition, the general consensus has been that insects hold the potential to positively contribute within diets (Bukkens 1997; Rumpold and Schlüter 2013). In fact, insects have been employed as a means to enrich otherwise nutritionally poor diets (Banjo et al. 2006; Bodenheimer 1951; Kinyuru et al. 2009; Konyole et al. 2012; Santos Oliveira et al. 1976; Tao and Li 2018). However, even with growing data on their use as a supplemental or complementary food data on the nutritional composition of insects still reports large margins of variability (Dobermann et al. 2017; Payne et al. 2016); presumably due to variation in processing and rearing methods (Fombong et al. 2017; St-Hilaire et al. 2007).

The impact of rearing methods, specifically choice of feed, is of particular note as it has previously been assumed that insects can easily be efficiently reared on food-based bio-waste sources (Van Huis et al. 2013). However, while lab scale trials have shown some success with rearing insects on bio-wastes (Caparros Megido et al. 2016a, b; Miech et al. 2016; Ramos-Elorduy et al. 2002), commercial scale trials have proven unsuccessful with significant negative impact on overall survival and development (Dobermann et al. 2018; Lundy and Parrella 2015).

Additionally, research has shown that the feed conversion efficiency of insects on bio-waste sources was not significantly better than poultry (Dobermann et al. 2018; Oonincx et al. 2015a). The only insect to consistently thrive and efficiently convert bio-waste feed is *Hermetia illucens,* the black soldier fly (Banks et al. 2014; Newton et al. 2005; Oonincx et al. 2015b; Sheppard et al. 1994).

While research has moved forward in leaps and bounds there are still large gaps in understanding with regard to the nutritional quality of insects and how to efficiently utilise them within existing food systems. A series of systematic trials are needed to elucidate the full practical potential of insects as food.

'Insects to Feed the World'

The FAO Paper (2013) and the International Conference 'Insects to Feed the World'

Many international organizations are trying to address the problem of malnutrition around the world, among these the Food and Agricultural Organisation (FAO) is promoting and raising awareness on the use of insects as a food and feed source since 2003 (Berg et al. 2017). As a consequence, over the past years, several projects related to 'sustainable insect harvesting' to benefit the communities were developed, both in countries where entomophagy is accepted/traditionally-known and in countries not familiar with this eating habit (Raheem et al. 2018).

In 2013, the FAO released a report 'Edible insects: future prospects for food and feed security' in FAO Forestry Paper 171, (Rome, 2013) (Van Huis et al. 2013) containing the State-of-the-Art on the topic of edible insects as human and animal food source. The work is a collaboration between FAO's Forestry Department in collaboration and the Laboratory of Entomology at Wageningen University in the Netherlands. The report is composed by 15 chapters, enriched by bibliographic references, suggestions for further reading, explanatory boxes, figures and tables to make the topic of edible insects understandable and available for stakeholders as well the general public.

It reports a wide range of research and information on the breeding, the transformation, the conservation and the consumption of insects, and how this practice can contribute to food availability. Also, the prospects and the opportunities of commercial scale breeding are examined to improve the production of food for animals and humans. The document can be downloaded for free (http://www.fao.org/edible-insects).

A year later, on May 14–17, 2014, the first international conference on insects for food and feed "Insects to feed the world" was held at the Conference centre De Reehorst in Ede (Wageningen), the Netherlands. At this event, which received a wide global media attention, over 450 participants from 45 nations participated to discuss on the latest research, business and policy making key aspects in this sector, such as collection, production, processing, nutrition, food safety, legislation and policy, environmental issues, insects as feed, marketing, consumer attitudes and gastronomy, related to insects (Van Huis et al. 2015). The conference can be considered as one of the first milestones in the recognition of the professional insect industry and highlighted the need to produce animal proteins in a more sustainable way.

The outcomes clearly reported how this new developing sector is growing and the dynamic nature of using insects for food and feed worldwide. Among the major challenges emerged was how to develop awareness with the general public to promote insects as healthy, sustainable and tasted food.

In 2018, the second International Conference 'Insects to Feed the World' (IFW 2018) was held in Wuhan, China P.R. and was attended by 278 individuals from 40 countries. Most of the focus regarded the black soldier fly, *Hermetia illucens* (L.) (Diptera: Stratiomyidae) used in the feed industry. Regarding the food industry, which is rapidly growing in some Western countries, insect-based products using cricket flour seem to be the most popular. Compared to the first conference, a series of novel topics were covered such as insect physiology, microbiome, mechanisation, economics, and quality assurance of insect mass production facilities, historical perspectives of insects as food and feed (Tomberlin et al. 2018).

The Response in Europe

The Media

In the last years, besides the increasing interest in edible insects from research institutes and the food industry, entomophagy has quickly gained a particular attention in both local and national media in Europe, especially in Northern countries where there has been an ample media coverage, sometimes becoming a trend topic (Gmuer et al. 2016; Menozzi et al. 2017; Piha et al. 2016).

In particular, in a study carried out by Myers and Pettigrew (2018), they reported how documentaries, reality television programs, news reports, and books are the most familiar channels to communicate on insect eating. However, as indicated by Dobermann et al. (2017) some of these tv shows (i.e. Fear Factor' and 'I'm A Celebrity...Get Me Out Of Here!') spread the idea of insects as inedible and disgust.

Nevertheless, after the events organized during the Universal Exposition 2015 in Milan, Italy, and the release of the FAO report 'Edible insects: future prospects for food and feed security' (2013), a great number of news on mass media channel (i.e. the BBC and The Sun in UK) started to promote insects as the "food of the future" (Shelomi, 2016). In particular, mass media highlighted the nutrition profile of insects (e.g. high-protein food) and the ecologically sustainable and pro-climate consumption which start changing the perception on eating insects among the general public (Caparros Megido et al. 2018; Piha et al. 2016; Sogari 2015).

Previous studies have suggested that the knowledge of and past exposure to entomophagy is strictly related with an increased likelihood to eat insects (Caparros Megido et al. 2016a; b; Hartmann and Siegrist 2017; Sogari et al. 2018). As remarked by Van Huis et al. 2015, p. 4), one of the major areas which requires urgent attention is *"improving communication, outreach strategies and messaging to the public at large ... on the potential, opportunities and acceptability of insects to contribute to a more sustainable and socially more equitable global food security"*.

The Research Community

A simple bibliographic search on the Scopus abstract and citation database using the words "edible insect" AND "entomophagy" reports approximately one hundred and fifty scientific manuscripts. A quick analysis shows that about 10 papers were published between 1995 and 2003 mainly by biologists and entomologists and that only two of them were published by European authors. The last manuscripts from 2003 show an awareness of the "disgusting" side of edible insects for westerners which potentially led to a lack of interest for insects as food. A few years later, the impact of livestock activities on all aspects of the environment (including air and climate change, land and soil, water and biodiversity) became a well-established concern and the need for environmentally friendly protein production could explain a new-found interest for edible insects. This interest was expressed by a renewal in edible insect studies (17 manuscripts over this period) but, still, a small interest from European research laboratories (5 European teams involved). In 2013, following the publication of the FAO report by van Huis et al. (2013), there was a significant increase in the number of publications each year (about ten in 2013 and 2014, about twenty in 2015 and 2016) reaching more than twenty-five publications in 2017 and in 2018. Edible insects, which have been of interest to just a small subset of biologists and entomologists in the past, are currently attracting researchers from a wider range of disciplines, from Food Quality or Safety Sciences (including micro-bial, chemical, toxicological, allergology or food formulation laboratories) to Social, Psychological, Environmental, Marketing or Economic Sciences. Moreover, European teams are actively involved in research projects related to edible insects: more than half of all publications in 2018 are conducted by European laboratories. The association of these different research areas through interdisciplinary projects is probably the key to the development of insects as food as an emerging sector in Europe.

Entrepreneurs

For those who look at the world of insects and entomophagy from a professional point of view, the FAO has produced a very useful tool. This tool is a Stakeholder Directory that contains references to the individuals involved at various levels - pri-vate, institutional, university, non-governmental, communication - in the world of entomophagy. This helps to communicate and disseminate information and give the possibility to start networks to collaborate on all aspects of insects as feed and food.

In addition to this, in the last years there have been several social events and conference around Europe (i.e. Insects as food and feed - an interdisciplinary work-shop in 2015, Oxford, UK; Insecta in 2017, Berlin, Germany; and others) and new platforms have been established to share knowledge between operators in this new sector (i.e. the International Platform of Insects for Food and Feed (IPIFF) and the EAAP Study Commission Insects) (Veldkamp and Eilenberg 2018). However, in Europe this sector is still represented in its vast majority by small and medium-size enterprises, mainly start-up companies (Derrien and Boccuni 2018).

Legislation

The production of insects for food is covered by (EU) No. 178/2002 (2002) on human safety (CE No 178/2002 2002). Regulation 178/2002 lays down the general principles and requirements of food law, establishes the European Food Safety Authority (EFSA) and lays down procedures in matters of food safety (Caparros Megido et al. 2015). In 2013, the most relevant regulation of the legislation with regard to the suitability of insects for human consumption is the Novel Foods Regulation (EC) No. 258/97 (CE No 258/97 1997). This means that foods that have not been consumed to a significant degree in the European Union before 15 May 1997 must undergo a risk assessment before being marketed (Caparros Megido et al. 2015). Within the different categories proposed, insects fall into category (e): foods and food ingredients consisting of or isolated from plants and food ingredi-ents isolated from animals [...]. This regulation was very ambiguous, in terms of the interpretation of the "significant degree" of consumption as well as "ingredients isolated from animals" (Caparros Megido et al. 2015; Lotta 2017).

Consequently, two different approaches were adopted by European Union (EU) Member states: (1) some of them (e.g. Italy or Spain) considered whole insects and their part as novel food while (2) other members (Belgium, Denmark, United Kingdom (UK) and The Netherlands) considered whole insects and their parts out of the scope of the regulation, while substances isolated from insects (e.g. proteins or lipids) were considered as novel foods (Lotta 2017). Due to these ambiguities, the European Parliament produced a new novel food regulation: (EU) No. 2015/2283 (CE No 2015/2283 2015). This new regulation went into effect in January 1, 2018 and they repealed and replaced Regulation (EC) No. 258/97. Consequently, from 1 January 2018, all insect food operators must fulfil a novel food application that is specific to their products (type of species, time of harvesting, substrate used, methods for farming and processing, etc) (Lotta 2017). Finally, insects or insect-based products that are legally on the market on 1 January 2018 may continue to be placed on the market until 2 January 2020 if a novel food application is submitted by 1 January 2019 (Lotta 2017).

The Response in the United States

The Media

There has been growing momentum and interest in the area of Insects as Food in the United States. This can be observed with the increase in publications in the popular press addressing edible insects. A Google Search of "Edible Insects in the United States" with a term limit of 2008–2018 resulted in 11,200 individual articles and blogs featuring edible insects. A Google search of "Entomophagy in the United States" with the same 10-year term resulted in 844 articles (Google, 2018). This is not surprising however, since the term entomophagy is not as commonly used by the general public when referring to the human consumption of insects.

The Research Community

In the United States in the last ten years, two companies in the Insects as Food industry have utilized Small Business Innovation Research Program (SBIR) through the United States Department of Agriculture (USDA) National Institute of Food and Agriculture (NIFA). SBIR grants are specifically for new, unusual and high impact technologies, which insects as an alternative protein source certainly fall under. Insect Agriculture in the United States has the potential to create more agricultural jobs in both rural and urban communities on a global research scale. The SBIR grants received have been utilized for research and development of processing and manufacturing technologies, not the farming of the insects. Insect farmers to date have not received any support from the federal government. Research in the insects as food space has involved multiple disciplines in the United States including: Entomology, Anthropology, Food Science and Nutrition, English, Philosophy, Psychology, Business, Journalism, Sustainability, Marketing, Veterinary Medicine, Human Nutrition, Environmental Education, Youth Development, Culinary Arts, Epidemiology, Engineering.

Entrepreneurs

Startup companies and restaurants using insects as a food ingredient, as well as insect farms (mostly cricket and mealworm) rearing insects for human consumption is gaining traction. There are currently approximately twenty large and small scale insect farms for human consumption in the United States.

Trade Organization

The 2016 creation of the North American Coalition for Insect Agriculture (NACIA) at the *Eating Insects Detroit* Conference indicates a unified vision and voice for the insects as food industry in North America. "The mission of the North American Coalition for Insect Agriculture is to foster collaboration amongst stakeholders and create a consolidated voice to encourage the positive growth of insects as both feed and food" (NACIA.org, 2018). A subsequent conference solely dedicated to Insects as Food, *Eating Insects Athens (EIA)*, indicates a surge in culinary interest, research, businesses, and education in response to insects as food in the United States. *EIA* Conference highlights included:

- 40 speakers
- 3 Keynotes – Jack Armstrong (Fluker Farms), Dr. Julie Lesnik (Department of Anthropology Wayne State University), Pat Crowley (Chapul).
- Distinguished Achievement Awards for Florence Dunkel (Department of Plant Sciences and Plant Pathology, Montana State University) and Craig Sheppard (Department of Entomology, University of Georgia)
- 100+ attendees representing a mix of industry, academia, and general interest.
- Curiosity Corner kickoff at Creature Comforts Brewery and Cine Independent Theater, bringing insect cuisine and education to the public
- The Buzz – 8 course tasting menu
- Vendor reception, open to public, with companies in human food, pet treats, feed, and consulting, plus a 12 insect bug buffet allowing everyone to learn about and taste a variety of insects

– NACIA Board of Directors and Membership Meeting
– Insect Artist market with 21 artists, 60+ works, and sales

NACIA also has a core mission to interact with sister trade organizations around the world. The 2018 Insects to Feed the World Conference in Wuhan, China marked the first time that members from the United States and NACIA met with their counterparts from other international entomophagy groups, including the International Platform of Insects for Food and Feed (IPIFF), The Asia Food and Feed Insect Association (AFFIA), and the Insect Protein Association of Australia (IPAA.) In a panel discussion with more than 300 participants from 40 countries the panellists from these four insect trade organizations discussed a road map for developing a robust and sustainable insect agriculture industry. In addition to sharing research on the benefits and practicalities of using insects as food, the group wanted set guidelines for industrial hygiene practices, certification of growers, consumer and producer education and the development of food standards for insects as part of the Codex Alimentarius, the international food standards maintained by the Food and Agriculture Organization of the United Nations (FAO) and the World Health Organization.

Conclusion

Insects have a history of being used as food in both Europe and the US. Traditional insect foods covered a range of species and processing techniques. However, until recently, these practices were scarce, extinct or in decline, and did not reach a mainstream audience.

Following an FAO report that emphasised the potential of insects as food, interest in edible insects has grown in both Europe and the US. In the commercial sector, insect farms and insect products are financially viable due to increasing consumer demand, and capital from investors has facilitated the emergence of many new insect food companies. In the academic sector, insects as food and feed are attracting interest from a range of disciplines and this has generated a rise in scientific studies on this topic.

As discussed in this chapter, there remain many gaps in both our understanding of insects as food and our optimization of farming and processing methods. Priorities for both commercial and academic stakeholders in this sector include an increased understanding of health impacts and safety of insect foods, and increased efficiency of farming systems, particularly those that use bio-waste as feed. Fortunately, these aims are within reach. Due to increasing investment in terms of both commercial application and scientific research funding, and due to the ongoing actions of multiple stakeholders, the emerging insect food sector in Europe and the US can look forward to a bright future.

References

Backwell LR, d'Errico F (2001) Evidence of termite foraging by Swartkrans early hominids. Proc Natl Acad Sci 98(4):1358–1363

Banjo AD, Lawal OA, Songonuga EA (2006) The nutritional value of fourteen species of edible insects in southwestern Nigeria. Afr J Biotechnol 5(1684–5315):298–301. https://doi.org/10.5897/AJB05.250

Banks IJ, Gibson WT, Cameron MM (2014) Growth rates of black soldier fly larvae fed on fresh human faeces and their implication for improving sanitation. Tropical Med Int Health 19(1):14–22. https://doi.org/10.1111/tmi.12228

Berg J, Wendin K, Langton M, Josell A, Davidsson F (2017) State of the art report insects as food and feed scholars research library. Ann Exp Biol 5(2):1–9

Besemer E (2018) MSU to host 30th annual Bug Buffet Feb. 20. MSU News Service. Available online: http://www.montana.edu/news/17451/msu-to-host-30th-annual-bug-buffet-feb-20

Bodenheimer FS (1951) Insects as human food. In: Insects as human food. Springer, Dordrecht, pp 7–38

Bukkens SGF (1997) The nutritional value of edible insects. Ecol Food Nutr 36(2–4):287–319

Campana MG, Robles García NM, Tuross N (2015) America's red gold: multiple lineages of cultivated cochineal in Mexico. Ecol Evol 5(3):607–617

Caparros Megido R, Alabi T, Larreché S et al (2015) Risks and valorization of insects in a food and feed context. Ann Soc Entomol Fr 51:215–258

Caparros Megido R, Gierts C, Blecker C, Brostaux Y, Haubruge É, Alabi T, Francis F (2016a) Consumer acceptance of insect-based alternative meat products in Western countries. Food Qual Prefer 52:237–243. https://doi.org/10.1016/j.foodqual.2016.05.004

Caparros Megido R, Alabi T, Nieus C, Blecker C, Danthine S, Bogaert J, Francis F (2016b) Optimisation of a cheap and residential small-scale production of edible crickets with local by-products as an alternative protein-rich human food source in Ratanakiri Province, Cambodia. J Sci Food Agric 96(2):627–632. https://doi.org/10.1002/jsfa.7133

Caparros Megido R, Haubruge É, Francis F (2018) Insects, the next European foodie craze? In: Edible insects in sustainable food systems. Springer International Publishing, Cham, pp 471–479. https://doi.org/10.1007/978-3-319-74011-9_30

CE No 178/2002 (2002) Regulation (EC) No 178/2002 of the European Parliament and of the Council of 28 January 2002 laying down the general principles and requirements of food law, establishing the European Food Safety Authority and laying down procedures in matters of food safety

CE No 2015/2283 (2015) Regulation (EU) 2015/2283 of the European parliament and the council of 25 November 2015 on novel foods, amending Regulation (EU) No 1169/2011 of the European Parliament and of the Council and repealing Regulation (EC) No 258/97 of the European Parliament and of the Council and Commission Regulation (EC) No 1852/2001

CE No 258/97 (1997) Regulation (EC) No 258/97 of the European Parliament and of the Council of 27 January 1997 concerning novel foods and novel food ingredients

Chapco W, Litzenberger G (2004) A DNA investigation into the mysterious disappearance of the Rocky Mountain grasshopper, mega-pest of the 1800s. Mol Phylogenet Evol 3:1055-7903/$ doi:10.1016/S1055-7903(03)00209-4

DeFoliart GR (1975) Insects as a source of protein. Bull Entomol Soc Amer 21(3):161–163

DeFoliart GR (1988) Editor's corner. The Food Insects Newsletter 1(1). Available online: http://labs.russell.wisc.edu/insectsasfood/files/2012/09/Volume_1_No1.pdf

DeFoliart GR (2002) The human use of insects as a food resource: a bibliographic account in progress. Available online: http://www.food-insects.com/

Derrien C, Boccuni A (2018) Current status of the insect producing industry in Europe. In: Edible insects in sustainable food systems. Springer International Publishing, Cham, pp 353–361. https://doi.org/10.1007/978-3-319-74011-9_21

Di Vittorio A (ed) (2006) An economic history of Europe. Routledge, UK.

Dobermann D, Swift JA, Field LM (2017) Opportunities and hurdles of edible insects for food and feed. British Nutr Foundation Nutr Bull 42:293–308

Dobermann D, Michaelson L, Field LM (2018) The effect of an initial high-quality feeding regime on the survival of *Gryllus bimaculatus* (black cricket) on bio-waste. J Insects Food Feed:1–8. https://doi.org/10.3920/JIFF2018.0024

Dunkel FV, Payne C (2016) Introduction to edible insects. In: Insects as sustainable food ingredients, Elsevier, USA. pp 1–27

Evans J, Alemu MH, Flore R, Frøst MB, Halloran A, Jensen AB, Payne C (2015) 'Entomophagy': an evolving terminology in need of review. Journal of Insects as Food and Feed, 1(4), 293–305

Evans J (2018) Fool Magazine, June 2018, pp.144–149

Fleming N (2016) The worm has turned: how British insect farms could spawn a food revolution. The Guardian, Fri 8th April 2016. Available online: https://www.theguardian.com/environment/2016/apr/08/the-worm-has-turned-how-british-insect-farms-could-spawn-a-food-revolution

Fombong FT, Van Der Borght M, Vanden Broeck J (2017) Influence of freeze-drying and oven-drying post blanching on the nutrient composition of the edible insect Ruspolia differens. Insects 8(3). https://doi.org/10.3390/insects8030102

Ghosh S, Jung C, Meyer-Rochow VB (2018) What governs selection and acceptance of edible insect species? In: Edible insects in sustainable food systems. Springer, Cham, pp 331–351

Gmuer A, Nuessli Guth J, Hartmann C, Siegrist M (2016) Effects of the degree of processing of insect ingredients in snacks on expected emotional experiences and willingness to eat. Food Qual Prefer 54:117–127. https://doi.org/10.1016/j.foodqual.2016.07.003

Goebel T, Hockett B, Adams KD, Rhode D, Graf K (2011) Climate, environment, and humans in North America's Great Basin during the Younger Dryas, 12,900–11,600 calendar years ago. Quat Int 242(2):479–501

Hartmann C, Siegrist M (2017) Consumer perception and behaviour regarding sustainable protein consumption: a systematic review. Trends Food Sci Technol 61:11–25

Hyman, V. (2013). Insect bites: cooking with cicadas, New Jersey's newest crop. NJ.com May 29 2013. Available online: https://www.nj.com/entertainment/dining/index.ssf/2013/05/cooking_cicadas.html

Kelemu S, Niassy S, Torto B, Fiaboe K, Affognon H, Tonnang H, Ekesi S (2015) African edible insects for food and feed: inventory, diversity, commonalities and contribution to food security. J Insects Food Feed 1(2):103–119

Kinyuru JN, Kenji GM, Njoroge MS (2009) Process development, nutrition and sensory qualities of wheat buns enriched with edible termites (Macrotermes subhylanus) from Lake Victoria region, Kenya. Afr J Food Agric Nutr Dev 9(8):1739–1750. https://doi.org/10.4314/ajfand.v9i8.48411

Konyole SO, Kinyuru JN, Owuor BO, Kenji GM, Onyango C, Estambale BB et al (2012) Acceptability of Amaranth grain-based nutritious complementary foods with Dagaa fish (Rastrineobola argentea) and edible termites (Macrotermes subhylanus) compared to corn soy blend plus among young children/mothers dyads in Western Kenya. J Food Res 1(3):111. https://doi.org/10.5539/jfr.v1n3p111

Krahl T, Fuhrmann H, Dimassi S (2016) Coloration of cereal-based products. In: Handbook on natural pigments in food and beverages, Woodhead Publishing, UK. pp 227–236

Lesnik JJ (2018) Edible insects and human evolution. University Press of Florida, FL, USA.

Lockwood JA, DeBrey LD (1990) A solution for the sudden and unexplained extinction of the Rocky Mountain grasshopper (Orthoptera: Acrididae). Environ Entomol 19(5):1194–1205

Lotta F (2017) Authorizing edible insects under the novel food regulation, conference reading it right on novel food, London, 27 September 2017. https://www.ifst.org/sites/default/files/edibleinsects.pdf

Lundy ME, Parrella MP (2015) Crickets are not a free lunch: protein capture from scalable organic side-streams via high-density populations of Acheta domesticus. PLoS One 10(4):1–12

Madsen DB (1989) A grasshopper in every pot. Nat Hist 89(July):22–25

Manditsera FA, Luning PA, Vincenzo F, Catriona ML (2018) The contribution of wild harvested edible insects (Eulepida mashona and Henicus whellani) to nutrition security in Zimbabwe. J Food Compos Anal, 75:17-25

Manunza L (2018) Casu Marzu: A Gastronomic Genealogy. In Edible Insects in Sustainable Food Systems (pp. 139–145). Springer, Cham.

Menozzi D, Sogari G, Veneziani M, Simoni E, Mora C (2017) Eating novel foods: an application of the theory of planned behaviour to predict the consumption of an insect-based product. Food Qual Prefer 59:27–34. https://doi.org/10.1016/j.foodqual.2017.02.001

Meyer-Rochow VB (1976) The use of insects as human food. Food Nutr Notes Rev 33(4):151–153

Meyer-Rochow VB (2010) Entomophagy and its impact on world cultures: the need for a multidisciplinary approach. Edible Forest Insects 6(2):23–36

Miech P, Berggren Å, Lindberg J, Chhay T, Khieu B, Jansson A (2016) Growth and survival of reared Cambodian field crickets (Teleogryllus testaceus) fed weeds, agricultural and food industry by-products. J Insects Food Feed 2(4):285–292. https://doi.org/10.3920/JIFF2016.0028

Mlček J, Adámková A, Adámek M, Borkovcová M, Bednářová M, Kouřimská L (2018) Selected nutritional values of field cricket (Gryllus assimilis) and its possible use as a human food

Müller A, Evans J, Payne CLR, Roberts R (2016) Entomophagy and power. J Insects Food Feed 2(2):121–136

Myers G, Pettigrew S (2018) A qualitative exploration of the factors underlying seniors' receptiveness to entomophagy. Food Res Int 103:163–169

Newton L, Sheppard C, Watson D (2005) Using the black soldier fly, Hermetia illucens, as a value-added tool for the management of swine manure. Animal and Poultry. Retrieved from http://www.urbantilth.org/wp-content/uploads/2008/09/soldierfly-swine-manure-management.pdf

Oonincx DG, Van Broekhoven S, Van Huis A, Van Loon JJ (2015a) Feed conversion, survival and development, and composition of four insect species on diets composed of food by-products. PLoS One 10(12):1–20

Oonincx DG, van Huis A, van Loon JJ (2015b) Nutrient utilisation by black soldier flies fed with chicken, pig, or cow manure. J Insects Food Feed 1(2):131–139. https://doi.org/10.3920/JIFF2014.0023

Pager H (1973) Rock paintings in southern Africa showing bees and honey hunting. Bee World 54(2):61–68

Pager H (1976) Cave paintings suggest honey hunting activities in Ice Age times. Bee World 57(1):9–14

Parasecoli F (2014) Al dente: a history of food in Italy. Reaktion Books, UK.

Payne CL, Van Itterbeeck J (2017) Ecosystem services from edible insects in agricultural systems: a review. Insects 8(1):24

Payne CLR, Scarborough P, Rayner M, Nonaka K (2016) A systematic review of nutrient composition data available for twelve commercially available edible insects, and comparison with reference values. Trends Food Sci Technol 47(March 2016):69–77

Piha S, Pohjanheimo T, Lähteenmäki-Uutela A, Křečková Z, Otterbring T (2016) The effects of consumer knowledge on the willingness to buy insect food: an exploratory cross-regional study in Northern and Central Europe. Food Qual Prefer. https://doi.org/10.1016/j.foodqual.2016.12.006

Raheem D, Carrascosa C, Oluwole OB, Nieuwland M, Saraiva A, Millán R, Raposo A (2018) Traditional consumption of and rearing edible insects in Africa, Asia and Europe. Crit Rev Food Sci Nutr. https://doi.org/10.1080/10408398.2018.1440191

Ramos-Elorduy J, González EA, Hernández AR, Pino JM (2002) Use of Tenebrio molitor (Coleoptera: Tenebrionidae) to recycle organic wastes and as feed for broiler chickens. J Econ Entomol 95(1):214–220. https://doi.org/10.1603/0022-0493-95.1.214

Rumpold BA, Schlüter OK (2013) Nutritional composition and safety aspects of edible insects. Mol Nutr Food Res 57(5):802–823

Santos Oliveira JF, De Carvalho JP, Bruno De Sousa RFX, Simão MM (1976) The nutritional value of four species of insects consumed in Angola. Ecol Food Nutr 5(2):91–97. https://doi.org/10.1080/03670244.1976.9990450

Sheppard D, Newton G, Thompson S (1994) A value added manure management system using the black soldier fly. Bioresource. Retrieved from http://www.sciencedirect.com/science/article/pii/0960852494901023

Shelomi M (2016) The meat of affliction: Insects and the future of food as seen in Expo 2015. Trends in Food Science & Technology, 56, 175–179

Sogari G (2015) Entomophagy and Italian consumers: an exploratory analysis. Prog Nutr 17(4):311–316

Sogari G, Menozzi D, Mora C (2018) The food Neophobia scale and young adults' intention to eat insect products. Int J Consum Stud 2018:1–9. https://doi.org/10.1111/ijcs.12485

St-Hilaire S, Cranfill K, McGuire MA, Mosley EE, Tomberlin JK, Newton L, Irving S (2007) Fish offal recycling by the black soldier fly produces a foodstuff high in omega-3 fatty acids. J World 38(2):309–313

Sutton MQ (1995) Archaeological aspects of insect use. J Archaeol Method Theory 2(3):253–298

Tao J, Li YO (2018) Edible insects as a means to address global malnutrition and food insecurity issues. Food Qual Saf 2(1):17–26. https://doi.org/10.1093/fqsafe/fyy001

Tomberlin JK, Zheng L, van Huis A (2018) Insects to feed the world conference 2018. J Insects Food Feed 4(2):75–76

Van Huis A (2003) Insects as food in sub-Saharan Africa. Int J Trop Insect Sci 23(3):163–185

Van Huis A, van Itterbeeck J, Klunder H, Mertens E et al (2013) Edible insects: future prospects for food and feed security. Food and Agriculture Organization of the United Nations, Rome, Italy.

Van Huis A, Dicke M, van Loon JJA (2015) Insects to feed the world. Insects Feed World Summary Rep 1:3–5. https://doi.org/10.3920/JIFF2015.x002

Van Huis A, van Itterbeeck J, Klunder H, Mertens E et al (2013b) Edible insects: future prospects for food and feed security. Food and Agriculture Organization of the United Nations

Veldkamp T, Eilenberg J (2018) Insects in European feed and food chains. J Insects Food Feed 4(3):143–145

Warner D (2006) Bug has a long, colourful history. Irish Examiner, May 8 2006. Available online: https://www.irishexaminer.com/lifestyle/outdoors/dick-warner/bug-has-a-long-colourful-history-2699.html

Yen AL (2009) Edible insects: traditional knowledge or western phobia? Entomol Res 39(5):289–298

Yen AL (2015) Insects as food and feed in the Asia Pacific region: current perspectives and future directions. J Insects Food Feed 1(1):33–55

Chapter 3
How to Measure Consumers Acceptance Towards Edible Insects? – A Scoping Review About Methodological Approaches

Giovanni Sogari, Davide Menozzi, Christina Hartmann, and Cristina Mora

Abstract In recent years there has been a growing number of studies analysing consumer acceptance, preferences, choices and willingness to pay for insects and insect-based products as food.

The aim of this chapter is to draw conclusions from existing literature published in scientific journals about the overall state of research activity on consumer attitude and behaviour towards entomophagy.

A scoping review was conducted by searching electronic databases for relevant articles using a determined key-terms search strategy. The starting dataset (n = 1366) was screened and analysed by the authors and a total of 102 articles were included in the review.

Findings highlight how a large number of researchers worldwide have investigated the potential drivers and benefits motivating consumers to accept insect and insect-based products as food as well as the main barriers that prevent individuals from consuming them.

The main themes that emerged are linked to: (1) the type of insect species and studies performing sensory tasting sessions, specifically those comparing specific species types (e.g. crickets, fried grasshoppers) to general/vague names (i.e. willingness to consume insects or insect-based product); (2) psycho-social and attitudinal variables like intention to eat, willingness to try, familiarity, food neophobia, emotional experiences, willingness to eat and overall entomophagy acceptance; (3)

G. Sogari (✉)
Department of Food and Drug, University of Parma, Parma, Italy

Charles H. Dyson School of Applied Economics and Management, Cornell University, Ithaca, NY, USA
e-mail: giovanni.sogari@unipr.it; http://www.unipr.it

D. Menozzi · C. Mora
Department of Food and Drug, University of Parma, Parma, Italy
e-mail: davide.menozzi@unipr.it; cristina.mora@unipr.it;
http://www.unipr.it

C. Hartmann
Consumer Behavior, ETH Zurich, Zurich, Switzerland
e-mail: christina.hartmann@hest.ethz.ch; http://www.cb.ethz.ch/

© Springer Nature Switzerland AG 2019   27
G. Sogari et al. (eds.), *Edible Insects in the Food Sector*,
https://doi.org/10.1007/978-3-030-22522-3_3

information treatment about entomophagy (e.g. benefits/risks of eating insects); (4) socio-demographic variables like differences in culture/country of origin (i.e. cross-country studies), gender, age and others.

A summary of the included records analysed as well as recommendations for future studies on how to develop research on consumer behaviour towards edible insect as food are explored.

Keywords Entomophagy · Behaviour · Novel food · Neophobia · Disgust

Introduction

After the approval of the new European Novel Food Regulation at the end of 2015 (Reg. (EC) 2283/2015), which applied starting from January 1, 2018, a growing interest in the consumer perspective has addressed the potential introduction in the market of edible insects and insect-based food products as an alternative protein source (Belluco et al. 2017; Hartmann and Siegrist 2017a; Sogari et al. 2019b; Tan et al. 2015).

The presence of high-quality protein and other elements (i.e. source of vitamins and minerals) (Rumpold and Schlüter 2013) as well as other potential nutritional, environmental, and economic benefits (Dobermann et al. 2017; van Huis and Oonincx 2017) make edible insect species one of the major potential future foods worldwide (Caparros Megido et al. 2018; van Huis 2013).

Usually novel food products are characterized by the use of new technologies which often create concerns and a climate of insecurity for consumers, mainly due to a lack of adequate information on the benefits and risks associated (Cardello and Wright 2010; Santeramo et al. 2018). Thereby consumer health and safety have always been considered among the most important factors for the acceptance of a novel food (Galati et al. 2019).

We know that edible insects have been consumed for millennials in many areas of the world (House 2018; van Huis et al. 2013) and, if properly managed, farmed, processed and consumed, can be considered safe for human consumption (Caparros Megido et al. 2018). However they still show low consumer acceptance as an alternative source to meat- and plant-based products (Schösler et al. 2012; Schouteten et al. 2016; Vanhonacker et al. 2013; Verbeke 2015).

This is mainly due to cultural appropriateness, social influence and also individual experiences which play a crucial role about what people accept and consider as food, even within a restricted geographic area (e.g. within the same country or region) (Hartmann et al. 2018; Caparros Megido et al. 2016; Verbeke 2015; Verneau et al. 2016). So far, most findings reveal how eating insects is not culturally accepted in countries where insects are not traditionally considered to be food (i.e. Western societies) and this type of unfamiliar food often evokes rejection and disgust (Evans et al. 2015; La Barbera et al. 2018; Menozzi et al. 2017a; Sogari, 2015).

Therefore, within a growing and competitive open global food market (Cunha et al. 2018), it becomes vital to investigate similarities and differences in consumer behaviour towards this novel food, both in Western and Eastern countries (Sogari et al. 2019a), taking into consideration cultural frameworks, eating habits, socio-demographics, and other attributes.

In recent years, there have been a growing number of studies published in scientific journals related to insects as food, mainly covering topics from safety, microbiology, farming, processing, nutritional properties, sensory properties and consumer acceptance (Belluco et al. 2013; Gao et al. 2018; Grau et al. 2017; Testa et al. 2017).

To our knowledge, a few systematic reviews have summarized the main results of original studies on consumers' approach to entomophagy (Hartmann and Siegrist 2017b; Mancini et al. in press), however without a focus and critical approach on the samples, study designs, methodologies, and techniques used to investigate this behaviour.

Thus, the aim of our study is to systematically search, select, and examine the existing scientific literature on consumer acceptance and behaviour regarding edible insects and insect-based food, describing and summarising the methodological approaches used so far. Identifying potential gaps and overlaps, we hope that results will inform the scientific community, policymakers and the private sector on how primary research on consumer behaviour can be advanced using and integrating different methods and techniques from sensory and consumer science.

Methodology

The Scoping Review Procedure

The methodology of this study consists of a scoping review which was performed in October and November 2018 (last updated on November 30th, 2018).

Scoping reviews often propose an exploratory research question aimed at mapping key concepts, types of evidence, and gaps in a defined and emerging area of evidence for which many different types of methods have been used (Colquhoun et al. 2014). This technique of literature review requires a series of steps to ensure a rigorous and transparent process and is used to map the relevant literature in the field of interest (Arksey and O'Malley 2005; Sargeant et al. 2006). It differs from a systematic or narrative review because this process requires analytical reinterpretation of the literature, and the research question is broader and more comprehensive so as to capture the full breadth of the literature (Arksey and O'Malley 2005; Levac et al. 2010).

Throughout the whole data screening process, the reviewers had an ongoing exchange of opinions and met several times to discuss and make decisions about the scoping review process (e.g. starting at the beginning stages with the development of the research question and the study inclusion and exclusion criteria).

Search Strategy

After the researchers identified and formulated the study question (i.e. what approaches have been used to study consumer behavior and eating insects?), a protocol was developed in order to include all the eligible research studies (the protocol is available online at Open Science Framework: https://osf.io/ev9ap). To achieve this, a set of search terms on relevant databases were identified.

For this study, a systematic search was carried out using nine main electronic bibliographic databases in the field of interest (Scopus; Web of Science; Food Science and Technology; PsycINFO; International Bibliography of the Social Sciences (IBSS); Business Source Complete; PubMed; EconLit; Agricola).

The search strings (Table 3.1) were tested and refined through several rounds of records identification and an exact search strategy for every bibliography database is available upon request to the authors. The articles were searched in the title, abstract or keywords of the record (TITLE-ABS-KEY) in the databases and only papers published in English were taken into consideration.

All original articles published in peer-reviewed journals from any time were considered with no restrictions regarding the date of publication. Thus, "grey literature" and unpublished research were excluded.

The researchers also identified some other relevant references that met the inclusion criteria (n = 7), not initially found in the online database search strategy, but rather identified by checking the reference lists of studies cited in the original papers.

Selection of Articles for Review and Data Extraction

The selected abstracts of these references were collected in Zotero, a reference citation management tool (Vanhecke 2008), and then imported into Covidence Online Software (https://www.covidence.org). The latter is a web-based systematic review program used to carry out a valid evidence synthesis process, through the identification of duplicates, tracking inclusion and exclusion criteria and the workflow for all data extraction (Kellermeyer et al. 2005). As a first step, the duplicates were removed (n = 789). Subsequently, as suggested by other researchers (Levac et al. 2010), two

Table 3.1 Search strings for the electronic search strategy

1. TITLE-ABS-KEY (consumer* OR buyer*)
2. TITLE-ABS-KEY (neo-phob* OR neophob* OR perception* OR attitude* OR accept* OR behav* OR disgust* OR prefer* OR choice OR choos* OR {willingness to eat} OR {willingness to try} OR {willingness to buy} OR {willingness to pay})
3. TITLE-ABS-KEY (entomophag* OR insect* OR insect-based*)

Note: The star "*" is used to consider also the words with the same common root (e.g. neophob* would pick up neophobia, neophobic, etc.)

researchers independently screened for relevance of the records idenfitied (n = 1366) by reading the title and/or abstract, and referring to the inclusion and exclusion criteria reported in Table 3.2. One of the main criteria was to incorporate only primary research studies; thus all reviews, opinions and editorials were dropped from our database. The final set of records was exported from Covidence and the relevant PDF files downloaded for analysis. Only studies judged to be relevant were retained for the full-text critical analysis (n = 203). The selected articles were further independently analysed by the authors of this chapter and only those meeting the inclusion criteria were included in the scoping review. As a result, the final set of records used in our scoping review included a total of 102 publications (the complete list of references is available online at Open Science Framework: https://osf.io/g5qjs). We used the same reporting guideline commonly known for systematic reviews—the PRISMA (Preferred Reporting Items for Systematic reviews and Meta-Analyses)- to record the number of excluded studies at each stage (Fig. 3.1).

All data of the selected records were compiled into an Excel spreadsheet table to facilitate interpretation and analysis. The publications were subjected to a "descriptive analytical" methodological evaluation to extract contextual or process-oriented information from each study (Colquhoun et al. 2014). Based on other reviews on consumer perception (e.g. Bryant and Barnett 2018; Hartmann and Siegrist 2017a), information was extracted according to (1) general characteristics of the study sample, including country of investigation, sample size, and year of data collection (2) study design, including whether it was a quantitative or qualitative study, whether information was provided, and other aspects (3) the main research question of the article, (4) the outcome measures (5) the type of edible insect and/or insect-based food studied (including whether a tasting session was provided, and (6) the main results and implications.

Following the analysis, the main characteristics of the different studies were summarised and interpreted.

Table 3.2 Inclusion and exclusion criteria of the scoping review

Inclusion criteria	Exclusion criteria
• All original article types, excluding dissertations/theses • Quantitative, qualitative and mixed study design • Studies focusing on edible insects (based on the definition of FAO report in 2013) and insect-based products • Full text paper published in peer-reviewed journals • Focus on consumer perception and behavior • Studies published in English • No date restrictions were applied	• Papers which do not present primary research studies (reviews, opinions and outlooks, conference papers and abstracts, commentaries, and editorials) • Focus on aspects not related to consumer behaviour and eating/accepting/buying edible insects (e.g. environmental impact of insect farming, nutritional characteristics, safety) • Studies not focusing on aspects of edible insects and insect-based products (e.g. field of entomology)

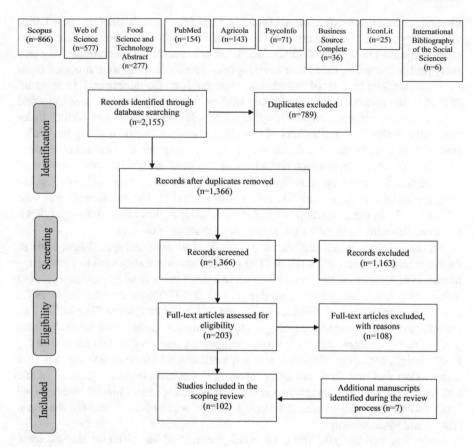

Fig. 3.1 Process for identifying and excluding the records based on Moher et al. (2009). This figure outlines the selection process. The search process identified 2155 articles from nine database sources, leaving 1366 after duplicates had been removed. After the titles and abstracts were screened for relevance, 203 articles remained. Another 108 articles were excluded after full-text review found that they did not meet the inclusion criteria. Finally seven articles were added from external sources reaching a total of 102 studies for the scoping review

Results and Discussion

The 102 papers containing relevant data for the scoping analysis have publication dates ranging between 2002 and 2019 (Fig. 3.2). Of those, 78% were published between 2016 and 2019 and report on data collected between 2014 and 2017. In 2014, the year after the publication of the FAO report "Edible insects: Future prospects for food and feed security" (van Huis et al. 2013), the number of publications increased substantially, and the trend has maintained since then.

A considerable number of papers (47% of the total) did not indicate the year of data collection, while four papers indicate multiple years of data collection.

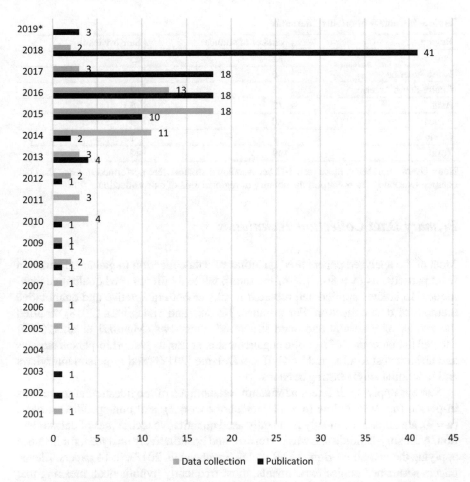

Fig. 3.2 Number of articles by year of publication and data collection. Note: overall number of studies n = 102. The year of data collection was not available in 48 studies; 4 papers indicate multiple years of data collection. *The few publications for year 2019 are due to the fact the electronic database search has been performed at the end of 2018 and only accepted publications by that date (November 30th 2018) have been considered

Most studies were carried out in Europe (n = 61), in particular, Western European countries (i.e. Italy n = 18, the Netherlands n = 13, Belgium n = 7, Switzerland n = 7 and Germany n = 6) (Table 3.3). Most of the remaining studies originated from countries where entomophagy is a traditional practice, such as in Africa (overall 20 studies, most of which were in Kenya n = 12) and Asia (overall 12, Japan n = 3, India n = 2, South Korea n = 2, Thailand n = 2, and China n = 2). Few original studies were carried out in North America (USA n = 6), Central and South America (mostly in Mexico n = 3), and Oceania (Australia n = 5). Overall, 7 studies were performed in more than one country (cross-cultural studies).

Table 3.3 Country of origin of the study

Region	Number of studies	Sensory testing
Europe	61	25
North America	6	2
Centre-South America	3	2
Asia	12	5
Africa	20	9
Oceania	7	1
Total	**109**	**44**

Note: Overall number is more than 102 because some studies were performed in more than one country. One study did not report the country or regional area of data collection

Primary Data Collection Techniques

Most of the analysed papers used quantitative data collection to gather information from participants (n = 80, 78% of the total), while 37 (36%) used qualitative techniques. 15 studies applied and reported results of both qualitative and quantitative methods of data collection. For instance, Nonaka and Yanagihara (2018) reported the results of a survey and semi-structured interviews, Menozzi et al. (2017a) showed the outcomes of a public engagement exercise, as well as in-person surveys and hedonic test results, and Le Goff and Delarue (2017) used in-person interviews and individual single tasting sessions.

Surveys applying different administration methods of the questionnaire, such as in-person (n = 18), online (n = 11), telephone (n = 2), and non-specified surveys (n = 9), are among the mostly frequently used quantitative techniques of data collection. A sensory-hedonic test was performed and quantitatively analysed, for instance applying the just-about-right – JAR scales (Pambo et al. 2018), in 14 papers. Eleven papers performed choice experiments, most frequently hypothetical, meaning that participants did not actually eat and/or purchase insects.

If we consider the qualitative studies, the most relevant data collection techniques were in-person interviews (both unstructured and semi-structured, n = 16) and focus groups (n = 10), often applied in the same study. Other less relevant qualitative techniques are public engagement activities and direct observation, including ethnographic studies. Even in qualitative studies, a tasting session often accompanied data collection. For instance, Tan et al. (2015) performed a cross-cultural qualitative study with focus groups exploring how cultural exposure and individual experience contributed to contrasting evaluations of insects as food.

Considering the sample, the main target population across studies was the general population (34%), university and undergraduate students and staff members (28%), and particular segments of consumers (including insect consumers, meat consumers, and general consumers, 15%). A lower number of studies examine specific population categories (e.g., mothers, children, caregivers, etc., 8%) or other convenience categories (e.g., visitors in cultural centres, etc., 8%). Finally, 7 studies did not indicate the type of sample (Table 3.4).

Table 3.4 Number of studies by target population

Type of sample	N
General public[a]	35
University, college students and staff	29
Consumers (e.g., insect consumers, meat consumers, and general consumers)[a]	15
Specific category (e.g., mothers, caregivers, rural households, children, etc.)	8
Other convenience (e.g., visitors of the cultural center, panellists, etc.)	8
Not available	7
Total	**102**

Note[a]: Samples of the general public and consumers are not necessarily representative of the target population (i.e. the probabilistic nature of the sample is explicitly cited in only four studies conducted with general population)

Table 3.5 Number of studies by sample size

Sample size (number of individuals)	N
<50	14
50–99	19
100–199	30
200–299	16
300–399	6
400–499	6
500–1000	15
>1000	8
Not available	4
Total	**118**

Note: Overall number is more than 102 because some publications considered more than one study/sample

Table 3.5 shows that 14 observations were retrieved from (mostly qualitative) studies with small sample sizes (i.e., from 15 to 50 individuals). A considerable number of studies (n = 30) were carried out with samples ranging from 100 and 190 individuals. 35 studies were performed with larger samples (300 or more individuals).

Studied Insect Species and Studies Performing Sensory Tasting Sessions

Overall, many different insect species were used in the selected studies; the most studied species were crickets (e.g., *Acheta domesticus* L., *Gryllodes sigillatus*, n = 32), mealworms (e.g., *Tenebrio molitor* L., n = 20), grasshoppers (n = 10), termites (e.g., soldiers of *Macrotermes falciger*, M. *natalensis* and M. *michaelseni*, n = 8), silkworms (n = 6) and locusts (e.g., *Locusta migratoria*, n = 5) (Table 3.6). 48 studies have investigated one single insect species, 12 studies analysed consumers' preferences and acceptance for two or three insect species, while 8 studies have considered more than three insect species. The remaining studies (n = 34) gave participants a general indication of "insects" without specifying any species.

Overall, 61 types of insects were used in tasting sessions (Table 3.6), most of which considered multiple (two or more) insect species in the same study. The insect species mostly used in tasting sessions were crickets (n = 23), mealworms (n = 13), termites (n = 5), grasshoppers (n = 4) and locusts (n = 4).

Considering both qualitative and quantitative studies, 42 studies out of 102 did actually perform some kind of sensory testing, two of them in more than one country (Lensvelt and Steenbekkers 2014; Tan et al. 2015). The tasting sessions were carried out in all geographic areas, independent from the actual availability of the insect-based products on the market (Table 3.3). The sensory properties most frequently studied were overall taste (n = 21), expectations of liking before tasting (n = 19), flavour, appearance and texture (Table 3.7).

Overall, 22 studies did not specify the insect species, and thus did not provide specific information to participants about the insect consumed as food (e.g., consumers' readiness to adopt insects as a meat substitute, in Verbeke (2015) or contained in the food (e.g., attitudes and intention to consume a novel food containing insect flour, in Menozzi et al. (2017b) (Table 3.8). We can assume that using the general term "insects" or a more specific species name will have an impact on the research outcome. The same can be said when showing pictures of insects as food rather than make participants imagine the situation.

Table 3.6 Number and type of insect species used in the selected studies and in the tasting sessions

Insect species	N. of studied insect species	N. of insect species used in tasting sessions
Crickets	32	23
Mealworms	20	13
Grasshoppers	10	4
Termites	8	5
Silkworms	6	2
Locusts	5	4
Bees	4	3
Ground ants	3	2
Caterpillars	3	1
Scorpions	2	0
Giant waterbugs	2	1
Cockroaches	2	1
Vespula spp.	2	1
Wasp larvae	2	0
Other	13	1
Total	**119**	**61**

Table 3.7 Number of studies investigating the sensory properties of insect-based food

Sensory property	N
Taste	21
Expected liking	19
Flavour	13
Appearance	13
Texture	11
Smell	9
Colour	9

Table 3.8 Type of insect food product used in the study

Type of insect food product	N
General insects as food (insect species not specified)	16
General insect-based food product (insect species not specified)	6
Specific type of insect food (visible insect)	22
Specific type of insect-based food (invisible insect[a])	37
Specific type of insect food (visible) AND insect-based food (invisible)	18
Not specified	2
So called insect-based food, but not containing insect	1
Total	**102**

Note[a]: insects are not visible, and the original taste of the whole insect taste is not particularly prominent and identifiable

Twenty-two studies included a specific type of insect food as a visible insect in the experimental study design (e.g., behavioural task using a chocolate with meal-worms on top, in Ammann et al. 2018), while 37 included a specific type of insect-based food where insects are not visible, and the original taste of the insect is not particularly prominent or identifiable (e.g., insect-based cookies, in Geipel et al. 2018). In that regard, the majority of studies tested consumer reactions to insect-based snack foods (e.g., cookies, biscuits, chips, chocolate bars, buns) and little is known about consumer acceptance of complete (ethnic) dishes with insects. Moreover, insects are often proposed as a potential meat replacer and products such as insect-based burgers or insect-based meatballs were tested in consumer studies. However, the tested products with insects as an ingredient (e.g. bruised insects, insect flour) only contained a small amount of insects; often less than 20 percent, which is probably due to the fact that insects are very expensive in Europe and North America and their processing properties are mediocre (House 2018). Interesting, but sparsely-tested approaches are insect-based sushi food and mixtures of traditional protein sources and insects (e.g. chicken-mealworm nuggets).

Often, the food samples used in the studies, with or without sensory testing, aimed at determining how preferences and taste/texture attribute evaluations varied between visible and non-visible insect products. This is the case of 18 studies where visible and invisible insect food products were presented to participants in the form of pictures or real foods (Table 3.8). For example Hartmann et al. (2015) analysed

Table 3.9 Psycho-social and attitudinal variables analysed

Personal factors	N
Intention to eat, willingness to eat, willingness to try	46
Sensory properties, liking	42
Acceptance, readiness to eat	32
Familiarity, past consumption	30
Attitudes towards insects, entomophagy	27
Food neophobia	25
Information, knowledge about entomophagy	24
Actual behaviour (intake, consumption, etc.)	24
Disgust and other negative emotions	23
Outcome expectations (environmental, health benefits, etc.)	20
Positive emotions (novelty, curiosity, etc.)	11
Willingness to pay	10
Attitudes towards ecology, environment	10
Preferences	10
Cultural background (incl. religion prohibition)	10
Subjective norms	8
Food choice motives	7
Risk perception	7
Attitudes towards health	6
Intention to but, willingness to buy/purchase	5
Perceived barriers (lack of practices, cultural)	5
Trust in information	2
Self-identity	1

attitudes towards and willingness to eat insect-based products in Germany and China, considering different kinds of insect-based processed (i.e., cookies based and chocolate chip cookies based on cricket flour) and unprocessed food items (i.e., deep-fried crickets and silkworms). In general, Western consumers reported higher willingness to eat the processed insect-based foods compared to the unprocessed foods, while more consumers in African and Asian countries are open to eating both invisible and visible insect-based food.

Psycho-Social and Attitudinal Variables Analysed

Intention, willingness to eat, and willingness to try are personal factors that have been examined most extensively based on the use of real products (e.g. in tasting sessions) as well as hypothetical experiments (e.g., using images) (Table 3.9). Forty-six papers analysed factors affecting the intention to eat or try insect-based food products, while sensory properties were analysed in 42 papers. Other personal

Table 3.10 Number of studies with information treatments, and significant effects on dependent variables

Information treatment	N	Effect
Negative individual information (sensory)	2	1
Neutral information	3	2
Positive individual information (health, nutritional, sensory/taste)	7	5
Positive societal information (environment, food security, economic development)	8	5
Information source (trust)	2	2

factors widely investigated were acceptance, readiness to eat these products, familiarity, attitudes towards entomophagy, food neophobia, and knowledge (both subjective and objective) about entomophagy. Likewise, an increasing number of studies investigated disgust sensitivity, which is a person's predisposition to experience disgust, as a predictor for insect food rejection. Actual behaviour (e.g., intake of insects, habitual consumption, etc.) was widely studied in developing countries, while disgust towards the insect product and other negative emotions connected with insect consumption were more frequently studied than positive ones (novelty, curiosity, etc.). Outcome expectations (e.g., environmental, health benefits, etc.) were investigated in 22 research studies.

Information Treatment

The role of information about the benefits and risks connected with the practice of eating insects has also appeared in several original studies (Table 3.10). These studies have investigated how willingness to try, preferences, choices and sensory properties varied between groups having been covered with different information treatments. In general, a larger number of studies have provided positive information about insect consumption, covering individual benefits (positive health effects, nutritional content, sensory, taste, etc.) as well as societal benefits (environment, food security, economic development, etc.). In many cases (5 out of 8) this information affects the dependent variables. A lower number of studies have included negative information (in particular about sensory property of the product). Also in these cases, the effect of information on the target variable has been significant.

Finally, two studies have considered trust in different information sources; the more significant effect has been found when information is provided by independent authorities (e.g., scientific researchers, persons using the product, the Government or relatives), instead of food producers or the media (see, e.g. Lensvelt and Steenbekkers 2014).

Table 3.11 Socio-
demographic variables
included in the study

Socio-demographic factors	N
Gender	18
Age	14
Place of origin (incl. Rural/Urban)	7
Educational level	4
Ethnicity	4
Socioeconomic position	2
Family size	2
Fitness practices (sport)	1

Socio-Demographic Variables

Finally, differences in culture, country of origin (i.e. cross-country studies) and other socio-demographics (i.e. age, gender) were often included in the studies as covariates (Table 3.11). Gender differences in the practice of eating insects were studied in 18 papers. In general, females have been found to be more sensitive to disgust than males and less willing to taste and eat insect food products. The relevance of age in explaining insect consumption was studied in 14 papers, where most of them found that insect consumption decreases with increasing age. However, other evidence has also been found. For instance, Tuccillo et al. (2018) found that young females are more disgusted than males, while no significant differences were observed between males and females among adults and the elderly. In some Japanese areas, insects hold some negative connotations especially among younger generations, while elderly people are more likely to have consumed edible insects in the past and are more open to purchasing insect products (Payne 2015). Other socio-demographic variables used as descriptors of insect consumption are place of origin, including the differences between rural and urban populations, and, to a lesser extent, educational level and ethnicity.

Conclusions, Future Trends and Research Needs

The aim of this review was to present a thematic synopsis of the work published, giving a concise but complete overview of the existing literature, without assessing the quality and results of every individual study.

Our work is one of the first reviews to use a comprehensive search strategy on consumer acceptance and insect eating implemented across multiple electronic citation databases without date restrictions.

Considering the research question was broad, a diversity of methodologies and techniques, used in sensory, social and economic science, have been found to investigate consumer behaviour in different contexts and populations. Sometimes the results across these studies are consistent, other times they are different and even controversial.

Many studies showed some kind of skepticism of Western consumers towards the introduction of edible insects in their diet, especially for visible and unprocessed insects, and less disgust for insect-based foods (Caparros Megido et al. 2014; Sogari et al. 2018; Vanhonacker et al. 2013). Most of the outcomes reveal that consumers socio-demographic factors and individual experiences (including past consumption) play an important role in acceptance (Hartmann and Siegrist 2016; Sogari et al. 2019b). In particular, the cultural framework can have a significant effect (Hartmann et al. 2015; Verneau et al. 2016), thus there is a need to carry out transnational research with a comparative approach to help companies to have a better understanding of consumer knowledge and openness in different markets. Overall, the main studies identified in this scoping review are geographically focused in Europe and the USA, however we believe that a special attention should also focus on Eastern countries like China given the large population and historic consumption of insects. It seems to be important to find out whether Western consumers are more open to insect preparation that mimics the traditional insect-based dishes such as those consumed in China, or whether the westernization of insect-based dishes is a prerequisite for acceptance. The appropriate flavouring of the insects plays a crucial role for their taste. However, there is a lack of research that directly compared acceptance ratings for traditional flavourings compared to westernized flavourings. Moreover, the meal context (e.g. snack food), meal setting (e.g. ethnic restaurant) and format (e.g. traditional meal composition) likely influence acceptance, but were rarely investigated in a structured way. For instance, testing consumers' reactions and emotional responses to insects in an ethnic restaurant setting might further shed light on contextual factors that influence the acceptance of novel foods such as insects.

It is likely that the study designs investigating consumer acceptance of edible insects will continue to change over the coming years as long as more of these products are available in the market. These developments will also enable the investigation of the effect of social influences on acceptance. Increased visiblity of insect products in restaurants, supermarkets and street food trucks is another crucial aspect. Not only how the social environment is a barrier or a driver of acceptance, but also how children and adolscents could be interested in eating such novel foods as insects needs to be investigated in the future. The majority of the studies investigated adults' reactions towards insects, but children might be an interesting target group as well. We observed how the willingness to try and preferences are sensitive to the information given to the participants. Therefore, marketing and insect-products advertising must be well-conceived. Highlighting certain characteristics of the insect-based product might unwillingly induce a disgust reaction and thus rejection. Moroever, there are almost no studies published that focused on repeat consumers. In the coming years, more studies which allow us to observe how repeat consumers of edible insects have changed their attitudes over time will give new insight on how a novel food can be accepted and, in the future, become a traditional food in our diet.

Lastly, in countries such as China, insects were part of the traditional cuisine, but seem to have slowly disappeared from people's plates. In order to better understand drivers and barriers of insect consumption in Europe, it might be worth investigating factors that are linked to insect acceptance and rejection in those countries where entomophagy was/is commonly practiced.

References

Ammann J, Hartmann C, Siegrist M (2018) Does food disgust sensitivity influence eating behaviour? Experimental validation of the food disgust scale. Food Qual Prefer 68:411–414. https://doi.org/10.1016/j.foodqual.2017.12.013

Arksey H, O'Malley L (2005) Scoping studies: towards a methodological framework. Int J Social Res Methodol Theory Pract 8(1):19–32. https://doi.org/10.1080/1364557032000119616

Belluco S, Losasso C, Maggioletti M, Alonzi CC, Paoletti MG, Ricci A (2013) Edible insects in a food safety and nutritional perspective: a critical review. Compr Rev Food Sci Food Saf 12(3):296–313. https://doi.org/10.1111/1541-4337.12014

Belluco S, Halloran A, Ricci A (2017) New protein sources and food legislation: the case of edible insects and EU law. Food Security 9(4):803–814. https://doi.org/10.1007/s12571-017-0704-0

Bryant C, Barnett J (2018) Consumer acceptance of cultured meat: a systematic review. Meat Sci 143:8–17. https://doi.org/10.1016/j.meatsci.2018.04.008

Caparros Megido R, Sablon L, Geuens M, Brostaux Y, Alabi T, Blecker C et al (2014) Edible insects acceptance by Belgian consumers: promising attitude for entomophagy development. J Sens Stud 29(1):14–20. https://doi.org/10.1111/joss.12077

Caparros Megido R, Gierts C, Blecker C, Brostaux Y, Haubruge E, Alabi T, Francis F (2016) Consumer acceptance of insect-based alternative meat products in Western countries. Food Qual Prefer 52:237–243. https://doi.org/10.1016/j.foodqual.2016.05.004

Caparros Megido R, Haubruge E, Francis F (2018) Insects, the next European foodie craze? In: Halloran A, Flore R, Vantomme P, Roos N (eds) Edible insects in sustainable food systems. Springer, Cham, pp 353–361. https://doi.org/10.1007/978-3-319-74011-9_21

Cardello AV, Wright AO (2010) Issues and methods in consumer-led development of foods processed by innovative technologies. In: Ahmed J, Ramaswamy HS, Kasapis S, Boye JI (eds) Novel food processing: effects on rheology and functional properties. CRC Press, Boca Raton

Colquhoun HL, Levac D, O'Brien KK, Straus S, Tricco AC, Perrier L et al (2014) Scoping reviews: time for clarity in definition, methods, and reporting. J Clin Epidemiol 67(12):1291–1294. https://doi.org/10.1016/J.JCLINEPI.2014.03.013

Cunha LM, Cabral D, Moura AP, de Almeida MDV (2018) Application of the Food Choice Questionnaire across cultures: systematic review of cross-cultural and single country studies. Food Qual Prefer 64:21–36. https://doi.org/10.1016/J.FOODQUAL.2017.10.007

Dobermann D, Swift JA, Field LM (2017) Opportunities and hurdles of edible insects for food and feed. Nutr Bull 42(4):293–308. https://doi.org/10.1111/nbu.12291

European Parliament and Council (2015) Regulation (EU) 2015/2283 of the European Parliament and of the Council of 25 November 2015 on novel foods, amending regulation (EU) No 1169/2011 of the European Parliament and of the Council and repealing regulation (EC) No 258/97 of the European Parliament and of the Council and Commission Regulation (EC) No 1852/2001. Off J Eur Union L327:1–22

Evans J, Alemu MH, Flore R, Frøst MB, Halloran A, Jensen AB et al (2015) 'Entomophagy': an evolving terminology in need of review. J Insects Food Feed 1(4):293–305. https://doi.org/10.3920/JIFF2015.0074

Galati A, Tulone A, Moavero P, Crescimanno M (2019) Consumer interest in information regarding novel food technologies in Italy: the case of irradiated foods. Food Res Int 119:291–296. https://doi.org/10.1016/j.foodres.2019.01.065

Gao Y, Wang D, Xu M-L, Shi S-S, Xiong J-F (2018) Toxicological characteristics of edible insects in China: a historical review. Food Chem Toxicol 119:237–251. https://doi.org/10.1016/J. FCT.2018.04.016

Geipel J, Hadjichristidis C, Klesse A (2018) Barriers to sustainable consumption attenuated by foreign language use. Nat Sustain 1:31–33. https://doi.org/10.1038/s41893-017-0005-9

Grau T, Vilcinskas A, Joop G (2017) Sustainable farming of the mealworm Tenebrio molitor for the production of food and feed. Z Naturforsch C (A Journal of Biosciences) 72(9–10):337–349. https://doi.org/10.1515/znc-2017-0033

Hartmann C, Siegrist M (2016) Becoming an insectivore: results of an experiment. Food Qual Prefer 51:118–122. https://doi.org/10.1016/j.foodqual.2016.03.003

Hartmann C, Siegrist M (2017a) Consumer perception and behaviour regarding sustainable protein consumption: a systematic review. Trends Food Sci Technol 61:11–25. https://doi. org/10.1016/j.tifs.2016.12.006

Hartmann C, Siegrist M (2017b) Insects as food: perception and acceptance findings from current research. Ernahrungs Umschau 64(3):M132–M138. https://doi.org/10.4455/eu.2017.010

Hartmann C, Shi J, Giusto A, Siegrist M (2015) The psychology of eating insects: a cross-cultural comparison between Germany and China. Food Qual Prefer 44:148–156. https://doi. org/10.1016/j.foodqual.2015.04.013

Hartmann C, Ruby MB, Schmidt P, Siegrist M (2018) Brave, health-conscious, and environmentally friendly: positive impressions of insect food product consumers. Food Qual Prefer 68:64–71. https://doi.org/10.1016/j.foodqual.2018.02.001

House J (2018) Insects are not 'the new sushi': theories of practice and the acceptance of novel foods. Soc Cult Geogr 9365:1–22. https://doi.org/10.1080/14649365.2018.1440320

Kellermeyer L, Harnke B, Knight S (2005) Covidence and Rayyan. J Med Libr Assoc 106(4):580–583. https://doi.org/10.5195/jmla.2018.513

La Barbera F, Verneau F, Amato M, Grunert K (2018) Understanding Westerners' disgust for the eating of insects: the role of food neophobia and implicit associations. Food Qual Prefer 64:120–125. https://doi.org/10.1016/j.foodqual.2017.10.002

Le Goff G, Delarue J (2017) Non-verbal evaluation of acceptance of insect-based products using a simple and holistic analysis of facial expressions. Food Qual Prefer 56:285–293. https://doi. org/10.1016/j.foodqual.2016.01.008

Lensvelt EJS, Steenbekkers LPA (2014) Exploring consumer acceptance of entomophagy: a survey and experiment in Australia and the Netherlands. Ecol Food Nutr 53(5):543–561. https:// doi.org/10.1080/03670244.2013.879865

Levac D, Colquhoun H, O'Brien KK (2010) Scoping studies: advancing the methodology. Implement Sci 5(69):1–18. https://doi.org/10.1017/cbo9780511814563.003

Mancini, S., Moruzzo, R., Riccioli, F., Paci, G. (in press). European consumers' readiness to adopt insects as food. A review. Food Res Int https://doi.org/10.1016/j.foodres.2019.01.041

Menozzi D, Sogari G, Veneziani M, Simoni E, Mora C (2017a) Eating novel foods: an application of the theory of planned behaviour to predict the consumption of an insect-based product. Food Qual Prefer 59:27–34. https://doi.org/10.1016/j.foodqual.2017.02.001

Menozzi D, Sogari G, Veneziani M, Simoni E, Mora C (2017b) Explaining the intention to consume an insect-based product: a cross-cultural comparison. In: Seyal AH, Al E (eds) Theory of planned behavior: new research. Nova Science Publishers, Inc, New York. pp 1–15

Moher D, Liberati A, Tetzlaff J, Altman DG, Group TP (2009) Preferred reporting items for systematic reviews and meta-analyses: the PRISMA statement. PLoS Med 6(7):e1000097. https:// doi.org/10.1371/journal.pmed.1000097

Nonaka K, Yanagihara H (2018) The contribution of high school students to succession and development of eating insect culture as a case of Vespula spp. J Insects Food Feed 4(Suppl. 1)

Pambo KO, Okello JJ, Mbeche RM, Kinyuru JN, Alemu MH (2018) The role of product information on consumer sensory evaluation, expectations, experiences and emotions of cricket-flour-containing buns. Food Res Int 206:532–541

Payne CLR (2015) Wild harvesting declines as pesticides and imports rise: the collection and consumption of insects in contemporary rural Japan. J Insects Food Feed 1(1):57–65

Rumpold BA, Schlüter OK (2013) Potential and challenges of insects as an innovative source for food and feed production. Innovative Food Sci Emerg Technol 17:1–11. https://doi.org/10.1016/J.IFSET.2012.11.005

Santeramo FG, Carlucci D, De Devitiis B, Seccia A, Stasi A, Viscecchia R, Nardone G (2018) Emerging trends in European food, diets and food industry. Food Res Int 104:39–47. https://doi.org/10.1016/j.foodres.2017.10.039

Sargeant JM, Rajic A, Read S, Ohlsson A (2006) The process of systematic review and its application in agri-food public-health. Prev Vet Med 75(3–4):141–151. https://doi.org/10.1016/j.prevetmed.2006.03.002

Schösler H, De Boer J, Boersema JJ (2012) Can we cut out the meat of the dish? Constructing consumer-oriented pathways towards meat substitution. Appetite 58(1):39–47. https://doi.org/10.1016/j.appet.2011.09.009

Schouteten JJ, De Steur H, De Pelsmaeker S, Lagast S, Juvinal JG, De Bourdeaudhuij I et al (2016) Emotional and sensory profiling of insect-, plant-and meat-based burgers under blind, expected and informed conditions. Food Qual Prefer 52:27e31

Sogari G (2015) Entomophagy and Italian consumers: an exploratory analysis. Prog Nutr 17(4):311–316

Sogari G, Menozzi D, Mora C (2018) Sensory-liking expectations and perceptions of processed and unprocessed insect products. Int J Food Syst Dyn 9(4):314–320. https://doi.org/10.18461/ijfsd.v9i4.942

Sogari G, Liu A, Li J (2019a) Understanding edible insects as food in Western and Eastern Societies. In: Bogueva D, Marinova D, Raphaely T, Schmidinger K (eds) Environmental, health, and business opportunities in the new meat alternatives market. IGI Global, Hershey, pp 166–181. https://doi.org/10.4018/978-1-5225-7350-0.ch009

Sogari G, Menozzi D, Mora C (2019b) The food neophobia scale and young adults' intention to eat insect products. Int J Consum Stud 43:68–76. https://doi.org/10.1111/ijcs.12485

Tan HSG, Fischer ARH, Tinchan P, Stieger M, Steenbekkers LPA, van Trijp HCM (2015) Insects as food: exploring cultural exposure and individual experience as determinants of acceptance. Food Qual Prefer 42:78–89. https://doi.org/10.1016/j.foodqual.2015.01.013

Testa M, Stillo M, Maffei G, Andriolo V, Gardois P, Zotti CM (2017) Ugly but tasty: a systematic review of possible human and animal health risks related to entomophagy. Crit Rev Food Sci Nutr 57(17):3747–3759. https://doi.org/10.1080/10408398.2016.1162766

Tuccillo F, Bonelli S, Torri L (2018) Disgust sensitivity towards insects and its relationship with demographic and cultural factors. J Insects Food Feed 4(Suppl. 1)

van Huis A (2013) Potential of insects as food and feed in assuring food security. Annu Rev Entomol 58:563–583. https://doi.org/10.1146/annurev-ento-120811-153704

van Huis A, Oonincx DGAB (2017) The environmental sustainability of insects as food and feed. A review. Agron Sustain Dev 37(5). https://doi.org/10.1007/s13593-017-0452-8

van Huis A, Van Itterbeeck J, Klunder H, Mertens E, Halloran A, Muir G, Vantomme P (2013) Edible insects. Future prospects for food and feed security. Food and Agriculture Organization of the United Nations, vol 171 Rome : Food and Agriculture Organization of the United Nations (FAO forestry paper 171) - ISBN 9789251075968

Vanhecke TE (2008) Zotero. J Med Libr Assoc 96(3):275–276. https://doi.org/10.3163/1536-5050.96.3.022

Vanhonacker F, Van Loo EJ, Gellynck X, Verbeke W (2013) Flemish consumer attitudes towards more sustainable food choices. Appetite 62:7. https://doi.org/10.1016/j.appet.2012.11.003

Verbeke W (2015) Profiling consumers who are ready to adopt insects as a meat substitute in a Western society. Food Qual Prefer 39:147–155. https://doi.org/10.1016/j.foodqual.2014.07.008

Verneau F, La Barbera F, Kolle S, Amato M, Del Giudice T, Grunert K (2016) The effect of communication and implicit associations on consuming insects: an experiment in Denmark and Italy. Appetite 106:30–36. https://doi.org/10.1016/J.APPET.2016.02.006

Chapter 4
Bugs on the Menu: Drivers and Barriers of Consumer Acceptance of Insects as Food

Christina Hartmann and Angela Bearth

Abstract Our daily food choices have a huge impact on the environment and on climate change. Animal based protein production in particular is very resource consuming. To satisfy the growing meat hunger in the world, alternative protein sources are needed that both have a smaller environmental impact and are readily accepted by consumers. Compared to beef and pork, plant and insect proteins can be produced more sustainably, although consumer acceptance may pose a particular challenge for the latter. In this chapter, we will explore Western consumers' acceptance of insects as food source and influencing factors. In particular, the role of emotional reactions towards insects, such as disgust and motivational barriers for the acceptance of insects as food will be discussed. Furthermore, the role of concepts taken from risk research, such as risk and benefit perception and trust, will be explored. Relevant characteristics of the insect product itself in terms of processing degree for consumers' willingness to eat will be highlighted. To further increase the sustainability of the insect production, food waste could be used as insect feed instead of more resource intensive feeds. The impact of different insect feeding styles on consumer acceptance and risk perception will be explained based on recent study results. Lastly, research gaps will be emphasized and strategies to overcome rejection of insects as food will be suggested.

Keywords Entomophagy · Consumer behavior · Novel food · Sustainable protein · Risk perception · Feed · Emotion

C. Hartmann (✉) · A. Bearth
Department of Health Science and Technology, Consumer Behavior, ETH Zurich,
Zurich, Switzerland
e-mail: christina.hartmann@hest.ethz.ch; angela.bearth@hest.ethz.ch; http://www.cb.ethz.ch/;
http://www.cb.ethz.ch/

© Springer Nature Switzerland AG 2019   45
G. Sogari et al. (eds.), *Edible Insects in the Food Sector*,
https://doi.org/10.1007/978-3-030-22522-3_4

Sustainable Protein Consumption

At the beginning of the nineteenth century, frequent meat consumption was an indicator of wealth and economic status (Bogueva et al. 2018; Teuteberg 1994). Today, meat is a cheap mass product, and meat consumption rates in Western countries are very high; the per-capita intake of meat (i.e. beef/veal, pork, poultry, sheep) in the European Union in 2016 was around 70 kg (retail weight) (OECD 2018). Therefore, intake rates of a substantial part of the population exceed dietary recommendations. To satisfy this hunger for meat in developed countries, there has been an evolution of animal husbandry systems and production methods that are based on efficiency and profit, where the environmental impacts and animals' natural needs and behavioral tendencies are often ignored (Rochlitz and Broom 2017; Steinfeld et al. 2006; Sumner et al. 2018). Our daily food choices have a huge impact on the environment. Production of meat has a much larger impact compared with the production of vegetable-based proteins, for example. To feed a growing population worldwide, researchers around the world are searching for new food technologies and resources. Interest in alternative protein sources of high nutritional value, such as edible insects, has increased remarkably in recent years (van Huis et al. 2013). New food technologies (e.g. cultured meat, genetic modification) and new food sources (e.g. insects) may help reduce the environmental impact of people's food behaviour (Bonny et al. 2015; Smetana et al. 2015). However, consumer acceptance of these new food sources is a challenge and a positive attitude toward such novel foods is a prerequisite for consumer acceptance.

Insects are in fact, depending on species, metamorphic stage and feeding style, rich in protein and essential amino acids, have a high vitamin and mineral content and low cholesterol concentrations compared to some meat-based animal products (Belluco et al. 2013; Verkerk et al. 2007). At the same time, their farming requires little water, space and their biomass conversation rate is better than that of most animals (van Huis et al. 2013). Both of these aspects, high nutritional value and smaller environmental footprint than traditional red meat production, makes insects particularly interesting as 'mini-livestock' (DeFoliart 1995) suitable for human and animal nutrition all over the world (van Huis et al. 2013). Thus, insects are a valuable food source in many parts of the world, including Africa, Latin America and Asia (van Huis et al. 2013). However, acceptance ratings for insects is low among Western consumers (Hartmann and Siegrist 2017), and even in countries where entomophagy is part of the traditional diet such as China, it has started to disappear (Chen et al. 2009; Hartmann et al. 2015). Shelomi (2016) highlighted this at the World Exhibition Expo 2015 in Milan, entomophagy appeared to be presented as an alternative for regions suffering from starvation, rather than as a modern, novel food option. But why is it so difficult to establish insects as a new protein source within Western dietary behaviour? In the following chapter we are going to introduce individual factors that influence Western consumers' acceptance of insects as food. We further highlight recent evidence about the importance of creating positive eating experiences and the potential of using role models and personality impressions for creating positive perceptions of insect food product consumers. Lastly, we present results from a new study about the impact of insect feeding style on consumer acceptance.

Individual Factors that Influence Acceptance of Insects as Food

Consumers who pay attention to the environmental impact of food choices reported a higher willingness to adopt insects as meat substitute (Verbeke 2015). However, insects are not a stable part of modern cuisines in most European countries and willingness to eat insects among the general population is rather low (Hartmann and Siegrist 2017). Especially women (Hartmann et al. 2015; Ruby et al. 2015; Schösler et al. 2012; Verbeke 2015) and those who value nutritional and health benefits of meat as well as those who focus on taste as a key component of meat quality, are less ready to adopt insects as meat replacement (Verbeke 2015). Another consistent finding across studies is that food neophobia, an individual's tendency to reject new and unfamiliar foods, is a negative correlate of willingness to eat insects (Hartmann et al. 2015; Hartmann and Siegrist 2016; Sogari et al. 2019; Tan et al. 2016a, b; Verbeke 2015). Food neophobia was once an important survival mechanism to prevent the ingestion of potentially poisonous substances (Pliner and Hobden 1992). It is associated with decreased levels of willingness to eat novel foods (Tuorila et al. 2001) and a barrier for the acceptance of insects as a new food source. Food rejection can be motivated by negative taste expectations and uncertainty about the origin of the product (Fallon and Rozin 1983; Tuorila et al. 1994). A typical reaction to things which people have learned are inedible or which are unfamiliar is disgust.

Disgust is a basic human emotion that prevents us from having contact with something that might be pathogenic, because it triggers behavioral avoidance of the stimulus. Thus, researchers consider it to be a component of the so called behavioral immune system (Terrizzi et al. 2013). People can vary in their tendency to react with disgust towards disgust elicitors that indicate the presence of pathogens such as certain odors (e.g. smell of decayed food) or visual cues (e.g. mold, runny nose). This disgust sensitivity was linked to a broad range of behavioral and attitudinal concepts in previous research. Disgust was by far the most frequently mentioned reason for rejecting eating insects in a study with Indian and US adults (Ruby et al. 2015). Results of that study further showed that those persons who scored high on the core disgust subscale – i.e. disgust based on a sense of offensiveness and threat of disease – (Haidt et al. 1994; revised by Olatunji et al. 2007) were less willing to eat insects. Food disgust sensitivity – disgust responsiveness towards certain food-related cues – and food neophobia together explained 37% of the variance in the willingness to eat insect products in another study with Swiss adults (Hartmann and Siegrist 2018). When considering hygiene, another important disgust domain, insects were long rather considered as an indicator for food contamination and a health risk than as a valuable food source in most Western societies (Kellert 1993; Lockwood 2013 p. 62; Looy et al. 2014). The presence of insects might even be considered as an indicator for low hygienic standards. Accordingly, an individual's susceptibility to be disgusted by poor food hygiene was a significant predictor for willingness to eat foods containing insects as an processed ingredient (Hartmann and Siegrist 2018). The majority of the aforementioned studies focused on a hypothetical willingness to eat insects on the one hand and disgust sensitivity on the other hand. But what about actual eating behavior? In a behavioral experiment conducted in our group, consumers were confronted with chocolate, which was decorated with dried mealworms. Again, participants' food disgust sensitivity strongly correlated with the amount consumed of the insect chocolate (Ammann et al. 2018). Therefore, (food) disgust sensitivity in previous research was not only linked to the hypothetical consumption of insect products, but also proved to be a significant predictor for actual eating behavior. These results nicely correspond with the finding by Sogari et al. (2019). In their experiment, they found out that intension to eat the insect products (processed and unprocessed) was strongly influenced by food neophobic tendencies, sociodemographic characteristics, sensory expectations and past exposure to insects as food; and intension to eat was highly correlated with actual eating of the insect products. Thus, they added further evidence for the link between the constructs discussed above and people's actual eating behavior.

Creating Positive Eating Experiences

Consumers' previous experiences with insects as food is one of the strongest predictor for its acceptance (Hartmann et al. 2015). For creating such positive experiences and overcoming initial reluctance, various researchers proposed different strategies. For instance, it was suggested that insects are prepared and presented with techniques that can usually be found in high gastronomy and by renaming them negative associations with the insect origin could be prevented (Deroy et al. 2015). Another proposed strategy is that insects are flavored with familiar spices (Caparros Megido et al. 2014) or incorporated into familiar dishes. Generally speaking, research showed that processed insects receive higher acceptance ratings than unprocessed insects, because evocative cues such as long legs that remind consumers on the insect origin of the food are not visible anymore (Hartmann et al. 2015). Consequently, a lot of recent research focused on insect products and dishes with grounded insects.

In one study with Swiss adults, study participants in the intervention group sampled insect-based tortilla chips while those in the control group ate traditional tortilla chips (Hartmann and Siegrist 2016). In both conditions, participants then indicated their willingness to eat unprocessed insects such as deep-fried silkworms and deep-fried crickets which were presented with a picture of those items. Results of that study showed that participants who ate the insect-chips before, reported a higher willingness to eat the unprocessed insects while simultaneously controlling for the impact of food neophobia, disgust and previous insect consumption which all had a significant influence. The fear of bad taste and negative textural properties play an important role in the rejection of unfamiliar foods (Pelchat and Pliner 1995). Thus, results of the aforementioned study support the notion that positive experiences with a processed insect food can lead to a higher willingness to consume the unprocessed counterparts as well (Hartmann and Siegrist 2016). Even though it might provoke false expectations concerning the taste of unprocessed insects, it might help to overcome the first hurdle to insect consumption and acceptance (Hartmann and Siegrist 2016).

Verneau et al. (2016) investigated with students from Denmark and Italy the effect of video-based information provision on consumers intension to eat insect products. Information about societal and individual benefits of introducing insect proteins into human diet were tested. Results showed that information provision did raise the intention to eat insects, and information about societal benefits appeared to be more stable over time than the effect of information on individual benefits. Noteworthy, intention was also reflected in participant's willingness to eat an insect-containing chocolate bar.

In another study (Looy and Wood 2006), educational presentations of "bug banquets" were carried out in order to alter negative attitudes towards insects as food. Students from different age levels were questioned concerning their attitudes towards insects before and after they had attended such a "bug banquet". The "bug banquet" included among other things whole cooked crickets and mealworms, roasted crickets and roasted seasoned mealworms and vegetable-based items such as spring rolls filled with carrots. Results showed that such an educational approach has subtle effects on attitudes. Nevertheless, the authors concluded that these occasions might help to increase familiarity with insects among consumers (Looy and Wood 2006). Unfortunately, even though results of these two studies suggested that information provision and creating insect-eating occasions could increasing willingness to eat, people who react with disgust towards insects are less likely to even go to such events (Hamerman 2016).

Such negative emotional reactions towards insects are a barrier for a successful market introduction. A study conducted in Switzerland investigated whether emotional reactions towards insects differs as a function of the processing degree of the insect ingredient (Gmuer et al. 2016). Again, researchers suggested to make insect products with grounded insects instead of unprocessed insects, because consumer acceptance is higher. The snacks used in the mentioned study were presented alongside pictures in an online survey. The insect snacks differed in their degree of processing of the insect ingredient: tortilla chips made of cricket flour, tortilla chips containing deep-fried cricket bits, a snack consisting of tortilla chips and deep-fried crickets, and deep-fried crickets alone. Respondents made 39 emotional evaluations, rated willingness to eat and expected liking of these snack products. Results showed that the insect snacks evoked various negative emotional expectations that went beyond expectations of disgust. Respondents did not expect positive emotional responses in the prospect of eating the snacks. Furthermore, expectations related to disgust/uneasiness and inertia/dissatisfaction were significant predictors of willingness to eat. The most negative evaluation received the mix product, which might point to food contamination associations. Overall, results highlighted that when it comes to the marketing of insects negative initial expectations need to be overcome such as disgust and dissatisfaction in the prospect of eating the foods, but also positive emotions should be generated (Gmuer et al. 2016).

A promising argument for the consumption of insects could be the high nutritional value of insect protein. Thus, health motivated consumers might be willing to eat insects for health reasons. However, it is not quite clear how consumers evaluate the healthiness of insect products anyway. One study compared healthiness perception of a menu containing either a vegetarian schnitzel, a pork schnitzel or an insect schnitzel (Hartmann et al. 2018). It turned out that consumers acknowledged higher nutritional benefits to the insect than to the pork menu, but compared to the vegetarian option, no difference was observed. The question arises why health motivated consumers then should consume insect products in the first place, when they do not see a nutritional benefit in insect consumption. In addition, various studies showed, however, that the taste of insects and insect products is not evaluated that positive, which makes it rather difficult to attract a stable consumer group (e.g. House in press; Schouteten et al. 2016). House (in press) conducted a qualitative study in the Netherlands where insect products are sold in some supermarkets as meat replacer (e.g. a schnitzel containing of 14% buffalo worms). The targeted market segment for these products were flexitarians who consciously try to reduce meat intake for environmental reasons. However, study participants' evaluations of the available products were rather negative. Problematic aspects mentioned were mediocre taste, high prices and low availability. These aspects make it rather difficult that insect products are preferred over comparable meat-replacer alternatives and make it rather unlikely that these products become a stable part of one's diet. In general, intention for repeated consumption of insect-based meat replacer was low (House in press) and these insect-based alternatives seem to suffer from the same problems like the traditional meat replacement products, mediocre taste and high price. In addition, it is not clear whether people eat such insect products in addition to meat or whether they in fact are eaten as meat replacers. When insects are just eaten as an additional source for protein, the goal to increase sustainability of food choices is not reached.

Creating Positive Impressions

Studies on impression management suggest a link between how people eat and how they are perceived by others. For example, a fictional woman was evaluated as more socially attractive when her meal was described as regularly sized and with a regular fat content as opposed to a high fat meal (Yantcheva and Brindal 2013). Moreover, Vartanian et al. (2007) concluded in their review of consumption stereotypes that people who eat "good" or low-fat foods are generally perceived as "better" people – that is, more attractive, intelligent, and conscientious. Prior research has also suggested that people who primarily consume plant proteins may be seen as more moral, more feminine, and more socially difficult than people who consume animal proteins (Ruby 2012; Ruby and Heine 2011). A person's food choices and eating behaviors are seen as reflections of lifestyle decisions, attitudes, and values, and this information is often used to form an impression of their personality. Prior results suggest that people who follow a vegetarian diet or consume meat alternatives, such as insects, might be perceived negatively. This would be an obstacle for increasing the sale of these products. In two experimental studies, both the shopping list method and a vignette approach were used to assess underlying impressions of these consumer groups. The aim of the first study was to explore how someone with insect-based or vegetarian burgers on their shopping list is perceived compared to someone purchasing beef burgers. Study participants (N = 598) were randomly assigned to one of three shopping list conditions and evaluated the owner of the list on 16 bipolar attributes (e.g., disciplined, health-conscious, popular). In the second study, a new set of participants (N = 617) was randomly assigned to one of three conditions. They read a short description about a hypothetical person who either chose a lunch menu with insect schnitzel, vegetarian schnitzel or pork schnitzel to elicit an evaluation of this person. The same personality attributes as in Study 1 were assessed. The results of both studies showed that consumers of insect and vegetarian products were perceived as more health-conscious, environmentally friendly, imaginative, brave, interesting, and knowledgeable than meat consumers. Both studies showed that insect consumers in Switzerland were evaluated positively. Given the relatively positive image of people who consume alternatives to traditional meat proteins identified in the present study, the social influence of people who visibly consume such products may be high. To increase the acceptance of insects as a food source, it is vital to recognize the importance of role models who demonstrate that eating alternative protein sources, such as insects, is a popular, environmentally friendly, and good-tasting option. This is especially important considering that social influences exert robust modelling effects on people's food intake (Spanos et al. 2015; Vartanian et al. 2007).

Insect Feeding Styles: Does It Matter for Acceptance?

In another recent experiment of our group, a question regarding the sustainability of insect breeding, was tackled: How relevant is the insect's menu for consumer perceptions? The environmental impact of edible insects depends largely on the insect feeding style (Smetana et al. 2016; van Huis and Oonincx 2017) and most insects species are highly efficient at bio-converting organic waste (Offenberg 2011). Life Cycle Assessments suggest that feeding insects with food waste, for example from restaurants and supermarkets, instead of rye, maize or soybean meal, could be a promising approach for more sustainable feed production (Offenberg 2011; Smetana et al. 2016). However, this raises some additional research questions, as consumers might be even more disgusted by insects that were fed food waste and thus, acceptability might be even lower. In an experiment we wanted to test this and presented consumers with a short introductory text on insects as high-quality and sustainable protein. Participants ($N = 613$, 52% female, $M = 45$, range: 20–69 years of age) were randomly distributed into four groups and received either a text describing that insects get fed with food waste from gastronomy, the supermarket, rye meal or other feedings stuff. Subsequently, they were asked to respond to four questions regarding their willingness-to-eat these insects, their risk and benefit perceptions and elicited disgust. Results suggest that consumers do not differentiate significantly between the insects fed with food waste from gastronomy, the supermarket, rye meal or other feedings stuff and exhibit similar values regarding willingness-to-eat, risk and benefit perceptions and disgust. Given the low overall willingness-to-eat ($M = 25.1$, $SD = 29.0$; range: 0–100) and high overall disgust ($M = 64.2$, $SD = 34.0$; range: 0–100), the difference in insect feeding style might not have been that salient for or important to consumers. Similarly, feeding style did not lead to different estimations of risk and benefit. In conclusion, feeding insects with food waste might not necessarily have detrimental effects on consumers' acceptance and might even have positive implications for consumers, as food waste is currently a much-discussed topic in need of consumer-oriented solutions (e.g., Hannibal and Vedlitz 2018; Stockli et al. 2018). Environmentally inclined consumers might perceive the possibility of converting food waste into high-quality protein as beneficial. This issue should be investigated further in future consumer studies that manipulate claims regarding sustainability of the edible insects with different feeds.

Concluding Remarks

Recent research has made much progress in understanding the psychology of eating insects. A lot of studies were published in the last couple of years that try to explain and provide strategies to overcome reluctance to eat insects in Western cultures. As daily food choices are embedded in cultural and social norms, food traditions and contexts, new food sources are difficult to establish as a stable part of Western diets. A lot of studies focused on individual factors that are linked to acceptance of insects as food. Factors such as food neophobic tendencies, previous experiences with eating insect, male gender and attention to the environmental impact of food influence consumers readiness to eat insects (Hartmann et al. 2015; Hartmann and Siegrist 2017; Sogari et al. 2019; Verbeke 2015). Another key driver for the acceptance of insects as food are of course positive first eating experiences, which increase future willingness to eat them. Especially those persons who seek for novelty and sensation in their diet are early adopters of food innovations like insects. Another underlying consumer motivation are environmental benefits of substituting traditional animal protein with insect protein. However, sustainability of insects as food highly depends on the insects' diet during breeding. Feeding insects with food waste substantially decreases their ecological footprint but might pose another challenge for consumer acceptance. Preliminary study results, however, showed that the feeding style of the insects is not of fundamental relevance for benefit and risk perception concerning insect consumption as well as evoked disgust and willingness to eat them. Next to low availability and price of insect products, which are barriers for repeated consumption, it is questionable whether meat enthusiasts can be convinced to substitute meat with insects without providing them with additional benefits. In that it might be necessary to give insects new functions (e.g. snack), new application forms (e.g. on a spit) and eating contexts (e.g. barbeque) instead of marketing them as a supplement to the traditional meat replacers. Overcoming lack of cultural appropriateness of insects for Western consumers by providing new functionality and eating contexts might pave the way for a wider acceptance.

References

Ammann J, Hartmann C, Siegrist M (2018) Does food disgust sensitivity influence eating behaviour? Experimental validation of the food disgust scale. Food Qual Prefer 68:411–414. https://doi.org/10.1016/j.foodqual.2017.12.013

Belluco S, Losasso C, Maggioletti M, Alonzi CC, Paoletti MG, Ricci A (2013) Edible insects in a food safety and nutritional perspective: a critical review. Compr Rev Food Sci Food Saf 12:296–313

Bogueva D, Marinova D, Phau I (2018) Is meat a luxury? In: Handbook of research on social marketing and its influence on animal origin food product consumption. IGI Global, Hershey, Pennsylvania, pp 172–186

Bonny SP, Gardner GE, Pethick DW, Hocquette J-F (2015) What is artificial meat and what does it mean for the future of the meat industry? J Integr Agric 14:255–263

Caparros Megido R et al (2014) Edible insects acceptance by Belgian consumers: promising attitude for entomophagy development. J Sens Stud 29:14–20

Chen X, Feng Y, Chen Z (2009) Common edible insects and their utilization in China. Entomol Res 39:299–303. https://doi.org/10.1111/j.1748-5967.2009.00237.x

DeFoliart GR (1995) Edible insects as minilivestock. Biodivers Conserv 4:306–321

Deroy O, Reade B, Spence C (2015) The insectivore's dilemma, and how to take the West out of it. Food Qual Prefer 44:44–55

Fallon AE, Rozin P (1983) The psychological bases of food rejections by humans. Ecol Food Nutr 13:15–26

Gmuer A, Nuessli Guth J, Hartmann C, Siegrist M (2016) Effects of the degree of processing of insect ingredients in snacks on expected emotional experiences and willingness to eat. Food Qual Prefer 54:117–127. https://doi.org/10.1016/j.foodqual.2016.07.003

Haidt J, McCauley C, Rozin P (1994) Individual differences in sensitivity to disgust: a scale sampling seven domains of disgust elicitors. Personal Individ Differ 16:701–713

Hamerman EJ (2016) Cooking and disgust sensitivity influence preference for attending insect-based food events. Appetite 96:319–326. https://doi.org/10.1016/j.appet.2015.09.029

Hannibal B, Vedlitz A (2018) Throwing it out: introducing a nexus perspective in examining citizen perceptions of organizational food waste in the US. Environ Sci Policy 88:63–71. https://doi.org/10.1016/j.envsci.2018.06.012

Hartmann C, Siegrist M (2016) Becoming an insectivore: results of an experiment. Food Qual Prefer 51:118–122. https://doi.org/10.1016/j.foodqual.2016.03.003

Hartmann C, Siegrist M (2017) Consumer perception and behaviour regarding sustainable protein consumption: a systematic review. Trends Food Sci Technol 61:11–25. https://doi.org/10.1016/j.tifs.2016.12.006

Hartmann C, Siegrist M (2018) Development and validation of the food disgust scale. Food Qual Prefer 63:38–50. https://doi.org/10.1016/j.foodqual.2017.07.013

Hartmann C, Shi J, Giusto A, Siegrist M (2015) The psychology of eating insects: a cross-cultural comparison between Germany and China. Food Qual Prefer 44:148–156. https://doi.org/10.1016/j.foodqual.2015.04.013

Hartmann C, Ruby MB, Schmidt P, Siegrist M (2018) Brave, health-conscious, and environmentally friendly: positive impressions of insect food product consumers. Food Qual Prefer 68:64–71

House J (in press) Insects are not 'the new sushi': theories of practice and the acceptance of novel foods. Soc Cult Geogr

Kellert SR (1993) Values and perceptions of invertebrates. Conserv Biol 7:845–855

Lockwood J (2013) The infested mind: why humans fear, loathe, and love insects. Oxford University Press, Oxford

Looy H, Wood JR (2006) Attitudes toward invertebrates: are educational "bug banquets" effective? J Environ Educ 37:37–48

Looy H, Dunkel FV, Wood JR (2014) How then shall we eat? Insect-eating attitudes and sustainable foodways. Agric Hum Values 31:131–141

Offenberg J (2011) Oecophylla smaragdina food conversion efficiency: prospects for ant farming. J Appl Entomol 135:575–581. https://doi.org/10.1111/j.1439-0418.2010.01588.x

Olatunji BO, Williams NL, Tolin DF, Abramowitz JS, Sawchuk CN, Lohr JM, Elwood LS (2007) The disgust scale: item analysis, factor structure, and suggestions for refinement. Psychol Assess 19:281–297

OECD (2018) Meat consumption (indicator) (Publication no. 10.1787/fa290fd0-en). Retrieved Accessed on 17 March 2018

Pelchat ML, Pliner P (1995) "Try it. You'll like it". Effects of information on willingness to try novel foods. Appetite 24:153–165

Pliner P, Hobden K (1992) Development of a scale to measure the trait of food neophobia in humans. Appetite 19:105–120

Rochlitz I, Broom D (2017) The welfare of ducks during foie gras production. Anim Welf 26:135–149

Ruby MB, Heine SJ (2011) Meat, morals, and masculinity. Appetite 56(2):447–450

Ruby MB (2012) Vegetarianism A blossoming field of study. Appetite 58(1):141–150

Ruby MB, Rozin P, Chan C (2015) Determinants of willingness to eat insects in the USA and India. J Insects Food Feed 1:215–225. https://doi.org/10.3920/jiff2015.0029

Schösler H, De Boer J, Boersema JJ (2012) Can we cut out the meat of the dish? Constructing consumer-oriented pathways towards meat substitution. Appetite 58:39–47

Schouteten JJ et al (2016) Emotional and sensory profiling of insect-, plant-and meat-based burgers under blind, expected and informed conditions. Food Qual Prefer 52:27–31

Shelomi M (2016) The meat of affliction: insects and the future of food as seen in Expo 2015. Trends Food Sci Technol 56:175–179

Smetana S, Mathys A, Knoch A, Heinz V (2015) Meat alternatives: life cycle assessment of most known meat substitutes. Int J Life Cycle Assess 20:1254–1267

Smetana S, Palanisamy M, Mathys A, Heinz V (2016) Sustainability of insect use for feed and food: life cycle assessment perspective. J Clean Prod 137:741–751. https://doi.org/10.1016/j.jclepro.2016.07.148

Sogari G, Menozzi D, Mora C (2019) The food neophobia scale and young adults' intention to eat insect products. Int J Consum Stud 43:68–76. https://doi.org/10.1111/ijcs.12485

Spanos S, Vartanian LR, Herman CP, Polivy J (2015) Personality, perceived appropriateness, and acknowledgement of social influences on food intake. Personality and Individual Differences 87:110–115

Steinfeld H, Gerber P, Wassenaar T, Castel V, Rosales M, De Haan C (2006) Livestock's long shadow FAO, Rome

Stockli S, Niklaus E, Dorn M (2018) Call for testing interventions to prevent consumer food waste. Resour Conserv Recycl 136:445–462. https://doi.org/10.1016/j.resconrec.2018.03.029

Sumner CL, von Keyserlingk MAG, Weary DM (2018) Perspectives of farmers and veterinarians concerning dairy cattle welfare. Anim Front 8:8–13. https://doi.org/10.1093/af/vfx006

Tan HSG, Fischer AR, van Trijp HC, Stieger M (2016a) Tasty but nasty? Exploring the role of sensory-liking and food appropriateness in the willingness to eat unusual novel foods like insects. Food Qual Prefer 48:293–302

Tan HSG, van den Berg E, Stieger M (2016b) The influence of product preparation, familiarity and individual traits on the consumer acceptance of insects as food. Food Qual Prefer 52:222–231

Terrizzi JA, Shook NJ, McDaniel MA (2013) The behavioral immune system and social conservatism: a meta-analysis. Evol Hum Behav 34:99–108

Teuteberg HJ (1994) Die historische Entwicklung des Fleischkonsums in Deutschland neu bewertet [The historical development of meat consumption in Germany newly evaluated]. In: Kluthe R, Kasper H (eds) Fleisch in der Ernährung [Meat for nutrition]. Georg Thieme Verlag, Stuttgart, pp 1–13

Tuorila H, Meiselman HL, Bell R, Cardello AV, Johnson W (1994) Role of sensory and cognitive information in the enhancement of certainty and linking for novel and familiar foods. Appetite 23:231–246

Tuorila H, Lähteenmäki L, Pohjalainen L, Lotti L (2001) Food neophobia among the Finns and related responses to familiar and unfamiliar foods. Food Qual Prefer 12:29–37

van Huis A, Oonincx D (2017) The environmental sustainability of insects as food and feed. A review. Agron Sustain Dev 37. https://doi.org/10.1007/s13593-017-0452-8

van Huis A, van Itterbeeck J, Klunder H, Mertens E, Halloran A, Muir G, Vantomme P (2013) Edible insects: future prospects for food and feed security (No. 171). Food and Agriculture Organization of the United Nations.

Vartanian LR, Herman CP, Polivy J (2007) Consumption stereotypes and impression management: How you are what you eat. Appetite 48(3):265–277

Verbeke W (2015) Profiling consumers who are ready to adopt insects as a meat substitute in a Western society. Food Qual Prefer 39:147–155

Verkerk M, Tramper J, Van Trijp J, Martens D (2007) Insect cells for human food. Biotechnol Adv 25:198–202

Verneau F, La Barbera F, Kolle S, Amato M, Del Giudice T, Grunert K (2016) The effect of communication and implicit associations on consuming insects: an experiment in Denmark and Italy. Appetite 106:30–36. https://doi.org/10.1016/j.appet.2016.02.006

Yantcheva B, Brindal E (2013) How much does what you eat matter? The potential role of meal size, fat content, and gender on ratings of desirability. Eating Behaviors 14(3):285–290

Chapter 5
Sensory and Consumer Perspectives on Edible Insects

Luís Miguel Cunha and José Carlos Ribeiro

Abstract Edible insects are part of traditional diets in several regions of the world such as Africa or Asia. However, in Western societies, the role of insects as food it is still not fully accepted mostly due to its rejection by consumers and to poor sensory properties of the insect-based food products. Rejection seems to be mainly regulated by disgust and food neophobia, hindering consumers' willingness to try edible insects or food products containing those as ingredients. In order to reverse this rejection, numerous strategies may be implemented, such as popularizing entomophagy, helping consumers to become familiarized with the concept or highlighting the nutritional and environmental advantages associated with eating insects. In addition to these rational discourses, it is of extreme importance to increase the sensory appeal of products containing edible insects and associate entomophagy with a positive sensory experience. However, there are several reports that the incorporation of insects in food products has a negative effect on their overall liking and yield sensory profiles associated with negative attributes. Therefore, even though rational strategies can be effective in profiling consumers ready to consume insects or increasing the number of consumers willing to try them, it is still necessary to improve the sensory properties of the insect-based foods to drive Western consumers into the adoption of edible insects in their regular diets.

Keywords Entomophagy · Acceptance · Neophobia · Insect-based food · Rejection · Disgust

L. M. Cunha (✉) · J. C. Ribeiro
GreenUPorto & LAQV/REQUIMTE, DGAOT, Faculdade de Ciências da
Universidade do Porto, Campus de Vairão, Vila do Conde, Portugal
e-mail: lmcunha@fc.up.pt

Introduction

Entomophagy (the practice of eating insects) is a part of traditional diets in several parts of the world (more specifically in Africa and Asia) with more than 2,000 species being consumed (van Huis et al. 2013). The main reasons for this practice are the nutritional value of insects (Rumpold and Schluter 2013), as replacement of conventional food sources when these are not available (Randrianandrasana and Berenbaum 2015) and also as consumers consider them tasty (Nonaka 2009). In the West, the situation is very different with insects not having been implemented in consumers' diets, but this situation has been changing in the last few years with a growing interest by the academic community and the food industry. While the legal uncertainty in the European Union (Belluco et al. 2017) and doubts surrounding food security (Ribeiro et al. 2018a; Vandeweyer et al. 2017) have hindered the implementation of entomophagy in the West, the main barrier is related to consumer acceptance. In this chapter entomophagy is analyzed from a consumer and sensory perspective. The main factors underlying entomophagy rejection and acceptance are reviewed. Strategies to increase entomophagy acceptance in the West are also revised. Lastly, studies assessing sensory properties of insect-containing products are assessed.

Factors Controlling Rejection of Entomophagy

Acceptance of entomophagy in the West is very low, with studies reporting that only 30–40% of Western consumers accept insects as food (Castro and Chambers IV 2018; Cunha et al. 2015). The main factors that control rejection of edible insects are food disgust and food neophobia (Cunha et al. 2015; Hartmann et al. 2015; Sogari et al. 2019; Verbeke 2015), although food disgust seems to play a greater role than food neophobia (Hartmann and Siegrist 2018; La Barbera et al. 2018).

Disgust is a primary emotion leading to the rejection of harmful substances that could be infected or that are sources of diseases (Chapman and Anderson 2012; Haidt et al. 1994). Food disgust can be provoked by distaste as in the case of bitter taste (Chapman and Anderson 2012) or by cues that symbolize hazardous items or the presence of pathogens such as alterations in colour, smell and taste (e.g. mould) in animal and non-animal foods provoked by spoilage and decay (Martins and Pliner 2006). Objects contacting a disgusting substance can also become a subsequent trigger of disgust due to contamination, as in the cases of food contaminated with human body fluids (Haidt et al. 1994). Moreover, reminders of animal origin or hygienic aspects of food preparation can also be cues for food disgust (Haidt et al. 1994; Martins and Pliner 2006). Nevertheless, the elicitors of disgust can greatly depend on the cultural and social environment of the individuals (Rozin and Haidt

2013), as seen in the case of insects which are regularly consumed in several regions of the World but provoke disgust in Western consumers. Thus, the elicitation of disgust provoked by insects must be based on learned associations between them and other basic sources of disgust (La Barbera et al. 2018). In fact, Western consumers do not view insects as food, associating them with vectors of disease, pests, spoiled food, dirtiness and lack of hygiene (Cunha et al. 2014; Looy et al. 2014; Rozin et al. 1986). Emphasized by the fact that some insects live in dirty natural habitats, rendering the entire category of insects disgusting (Chan 2019). Furthermore, when developing a general Food Disgust Scale, Hartmann and Siegrist (2018) report that the subscales that best predicted willingness to eat insect-based products were 'animal flesh' and 'poor hygiene', which are related to the animal-nature of the insects and their association with poor hygienic behaviour. It can also be argued that entomophagy disgust may not necessarily reflect a deep fear of contamination/diseases and is instead driven by social and cultural norms (Deroy et al. 2015; Jensen and Lieberoth 2019) and the perception of insects being inappropriate in existing food dishes and/or distasteful (Tan et al. 2016a). Nevertheless, works by Cunha et al. (2015) and Ruby et al. (2015) have shown that specific disgust at the thought of eating insects is a major determinant of entomophagy for consumers from diverse cultures, such as the American, the Indian, the Norwegian or the Portuguese.

Food neophobia is an established psychological construct that describes a person's tendency to reject or avoid eating unfamiliar foods or foods from other cultures. It can be greatly influenced by food-disgust sensitivity (Al-Shawaf et al. 2015), although these are two different psychological constructs (Hartmann and Siegrist 2018; La Barbera et al. 2018). This rejection can be a result of unknown origins or expected harmful consequences from consumption, especially if the food is from an unknown food origin (Martins and Pliner 2006). Food neophobia can also occur due to fear of bad sensory experiences caused by an unfamiliar food (Pelchat and Pliner 1995), insect-based products having low expectations of liking, sensory-profiling or even emotional-profiling (Gmuer et al. 2016; Schouteten et al. 2016; Tan et al. 2016b, 2017a, b).

Social and cultural norms can also negatively affect the acceptance of entomophagy, leading consumers to consider eating insects as culturally inappropriate (Myers and Pettigrew 2018). This effect seems to be greater in regions who have a strongly rooted more traditional food culture (Menozzi et al. 2017b; Sogari et al. 2019). Entomophagy can also be seen as a primitive or survival practice in poor countries, with insects only being consumed as an available alternative to other conventional food sources (Yen 2009), further contributing to the negative opinions of Western consumers. Social eating norms can also greatly influence individual consumer's willingness to eat insects when in collective tasting situations (Jensen and Lieberoth 2019; Sogari et al. 2017).

Profile of Consumers Willing to Eat Insects

The studies that have been performed on Western consumers' perception of ento-mophagy enable the identification of some characteristics of consumers who are more willing to accept edible insects. Gender seems to play a role, with several studies reporting that males have a higher acceptance of entomophagy (Hartmann et al. 2015; Verbeke 2015; Menozzi et al. 2017a; Woolf et al. 2019). This can be explained by the fact that in general, men are less sensitive to disgust than women and have a lower animal reminder disgust sensitivity (Hamerman 2016). Also, con-sumers who are aware of the nutritional and environmental impacts of their food choices present a higher acceptance of entomophagy (De Rosa et al. 2016; House 2016), due to the environmental and nutritional benefits that are associated with edible insects (van Huis et al. 2013).

Other potential consumers are those who seek new food experiences, as the nov-elty and curiosity of edible insects are main drivers of the intention to consume insects (House 2016; Menozzi et al. 2017b; Sogari et al. 2017). These results are in line with a study performed by Hartmann et al. (2018) in which consumers described buyers of insect burgers as more health-conscious, environmentally friendly, imagi-native, brave, interesting, knowledgeable, and athletic than buyers of beef burgers.

Children can also be considered as more easily accepting entomophagy, given that they might be adventurous enough to try products incorporating insects (Clarkson et al. 2018), although more studies are necessary to confirm this assumption. Nonetheless, introducing children to the concept of entomophagy can be a great way of guaranteeing that future generations will be more receptive of this practice, considering that disgust can be acquired at a very young age (Woolf et al. 2019).

Strategies to Improve Acceptability of Entomophagy

Rational Discourses

Several strategies can be implemented in order to increase Western consumers' acceptance of entomophagy. Popularizing entomophagy can be an effective initial strategy given that consumers familiar with the concept have a higher acceptance of edible insects (Cunha et al. 2014; Hartmann et al. 2015; Woolf et al. 2019). The popularization should also be attained by performing tasting sessions with insects given that the tasting of insects or products incorporating them can increase accep-tance of insects as food and make consumers more familiar with their sensory prop-erties (Menozzi et al. 2017b; Sogari et al. 2017, 2019). Familiarization can also have an effect on the types of insects that are more accepted, with Western consumers being more receptive to mealworms, crickets and grasshoppers; currently the most marketed species (Fischer and Steenbekkers 2018). This effect of familiarization is also evident in studies performed with consumers from regions in which insects are a common food (Hartmann et al. 2015; Tan et al. 2015), given that they were more willing to accept species present in their cuisine instead of those marketed in the West. This study highlights the cultural relativeness of food choice and the fact that the simple introduction to entomophagy will not be sufficient to drive Western consumers to include edible insects into their regular diets (Deroy et al. 2015).

It is also important to highlight the nutritional and environmental advantages that are associated with edible insects given that consumers who have some knowledge about them are more willing to consume insects (Verneau et al. 2016; Woolf et al. 2019). However, this kind of strategy might be insufficient when such foods are considered disgusting and hazardous (La Barbera et al. 2018). Focusing on the benefits of entomophagy will only be truly effective for consumers who are already prone to change in their dietary habits in accordance to their nutritional and/or environmental choices (Deroy et al. 2015; Hartmann et al. 2015; Verbeke 2015). Nonetheless, the current knowledge about the benefits of insect consumption is very low (Sogari et al. 2017; Woolf et al. 2019) and it is necessary to establish their role as a food source with high nutritional and environmental value in order to reach such consumers.

Sensory Appeal

Strategies should also be implemented to increase the sensory appeal of insects and insect-based food products, given that it is more effective than most communicational strategies (Hamerman 2016; Myers and Pettigrew 2018). Currently, improving the sensory appeal of insect-based foods is attained by associating them with known flavours and dishes while incorporating them in a processed, non-visible form (Gmuer et al. 2016; Hartmann et al. 2015; Hartmann and Siegrist 2016). This approach is already practised by the food industry, with many insect-based products already present in the market incorporating edible insects in a processed, non-visible form (with the most popular products being snacks such as cereal or protein bars and burgers/meatballs). This greater acceptance of processed insect-based foods is partly explained by the contrast with the visual appearance of whole insects that remind consumers of their animal-origin, increasing disgust reactions (Hartmann and Siegrist 2018). However, the consumption of processed insect-based food products upturns the consumers' willingness to consume unprocessed insects (Hartmann and Siegrist 2016). The same approach should also be applied when packaging products incorporating insects, given that the utilization of labelling with images of whole insects lowers the sensory appeal of the food product (de-Magistris et al. 2015). Nevertheless, it must be considered that masking the presence of food products containing insects can have the opposite effects on more "brave" and "enthusiastic" consumers, who are looking for a more genuine experience from entomophagy (Clarkson et al. 2018). Moreover, recent findings by Reinbold et al. (2018) show that when comparing different products with both visible and non-visible insects, consumers' liking was higher for ready-to-eat granolas with visible insects, while lower for other food, indicating that visible insects may improve acceptance regarded they are visually appropriate within the food matrix.

Repeated Consumption

The implementation of entomophagy in the West demands for the finding of a correct food categorization that makes sense for the consumer (Deroy et al. 2015). By guiding consumers to consider insect-containing products appropriate, one can greatly improve the willingness to eat and the expected sensory properties (Tan et al. 2016a, b, 2017a), thus enhancing the whole consumption experience. Usually, insects are presented as an alternative to meat due to their high protein content and more sustainable production. However, consumers expect meat substitutes to have similar sensory properties to meat (Deroy et al. 2015) and meat consumers are more driven by hedonic properties, being less willing to eat food sources who only present other advantages (Verbeke 2015). On the other hand, when consumers were given the opportunity to design an ideal insect-based product, they came up with such products as a convenient sweet snack, drink, or breakfast option, which were quick, healthy and a sustainable option for kids, fitness and/or health-oriented consumers(Clarkson et al. 2018). However, whether insects are categorized as a meat-alternative or are incorporated into snack-like products, it clear that edible insects are trying to be integrated into already-existing food practices, which is a striking difference from what occurred with the launch of other new food sources in the West, such as sushi (House 2018). With such positioning, insects become one of many options, having to compete with and to be subjected to a vast selection criteria (e.g. price, sensory properties, availability, convenience) and hindering their incorporation into the regular diet of consumers (House 2016; House 2018).

Sensory Properties of Insects and Insect-Containing Foods

Adding to the previously mentioned rational discourses, it is necessary to focus on the sensory properties of edible insects, considering that sensory evaluation is one of the key points evaluated by consumers when making food choices (Cunha et al. 2018). Moreover, developing tasty insect-based products, while associating them with positive gastronomic experiences, can lead to a lower incidence of disgust (La Barbera et al. 2018). Additionally, (poor) taste has been identified as one of the major reasons why consumers experiencing insect-containing products don't include them into their regular diets (House 2016). Consumers aren't willing to give up foods conveying positive experiences (sensory properties, price, availability, ability to fit in current diets) for others who only guarantee environmental, nutritional or health-related benefits (House 2016).

When assessing studies that have performed sensory evaluations of insects or insect-based products (Table 5.1), the majority of them have used crickets or mealworms (Farina 2017; Caparros Megido et al. 2014, 2016; Ribeiro et al. 2018b; Schouteten et al. 2016; Sogari et al. 2018; Tan et al. 2017b; Tao et al. 2017; Zhong 2017; Zielińska et al. 2018), although other species such as locusts (Tao et al. 2017; Zielińska et al. 2018), cockroaches (de Oliveira et al. 2017) or termites (Ogunlakin et al. 2018) have also been studied. Interestingly, in all the studies performed in

Table 5.1 List of studies assessing liking, acceptance and sensory properties of edible insects and insect-based food products

Article	Country	Species	Food products	Sensory evaluation methods	Results
Ribeiro et al. (2018b)	Portugal	Crickets *Acheta domesticus* and *Gryllodes sigillatus*	Cereal bars without cricket, with whole ground crickets or with defatted crickets.	Overall liking Food action rating scale Check-all-that-apply	Cereal bars incorporating defatted crickets and the control bar had very similar sensory evaluations and profiles. Cereal bars incorporating whole ground crickets had the lowest liking and acceptance scores and were associated with negative sensory attributes (e.g. 'unpleasant aftertaste', 'earthy', flavour, 'unpleasant' odour, 'rancid' or 'mouldy' odour, etc.)
Zhong (2017)	U.S.A	Cricket *Acheta domesticus*	Protein bar with cricket powder and market bar.	Overall liking Hedonic attributes evaluation Food action rating scale Paired-preference Triangle difference	The market bar had significantly higher evaluations values for flavour, aroma, texture, overall liking and FACT. 61% of participants preferred the market bar. 84% of the participants were able to correctly identify the cricket bar in a triangle test.
Zielińska et al. (2018)	Poland	Mealworm *Tenebrio molitor*, locusts *Schistocerca gregaria* and crickets *Gryllodes sigillatus*	Protein preparation and whole ground	Overall liking Hedonic attributes evaluation	Protein preparations had higher scores than the whole ground species, for all evaluations. Products with the best evaluation were protein preparation and whole ground *S. gregaria*. Whole ground *T. molitor* had the worst evaluations
Ogunlakin et al. (2018)	Nigeria	Termite *Macrotermes nigeriensis*	Biscuits incorporating wheat-termite flour	Overall liking Hedonic attributes evaluation	No differences in aroma and texture evaluation. Taste, crispiness, colour and overall liking were highest for control and biscuit with 95–5% wheat-termite flour
Sogari et al. (2018)	Italy	Cricket *Acheta domesticus*	Soft jelly with whole visible house cricket inside or with cricket flour	Sensory-liking expectation Hedonic attributes evaluation Paired preference Free listing of attributes	Majority of consumers preferred the product with the whole cricket. After tasting, evaluation of the taste of the jelly with whole cricket significantly increased and was superior to the taste score of the jelly with cricket flour. Products had different sensory profiles, especially with attributes concerning taste and texture.

(continued)

Table 5.1 (continued)

Article	Country	Species	Food products	Sensory evaluation methods	Results
Tao et al. (2017)	U.S.A	Cricket *Acheta domesticus* and locust *Locusta migratoria*	Extruded rice	Just-about-right questions regarding colour, aroma, flavour and mouthfeel. Overall liking	Rice with cricket flour had better attributes evaluation than rice with locust flour, especially concerning aroma and flavour. All samples were associated with a grainy texture. Extruded rice with cricket flour had higher overall liking than those with locust flour.
Farina (2017)	U.S.A	Cricket (not specified)	Broths with crickets that were frozen prior to cooking or still alive prior to cooking	Overall liking Hedonic evaluation of saltiness, sweetness, bitterness, sourness and umami flavour.	Differences were detected for overall liking and umami flavour, with broth made from frozen crickets having the highest scores.
de Oliveira et al. (2017)	Brazil	Cinereous cockroach *Nauphoeta cinerea*	Bread with 10& cockroach flour	Overall liking Hedonic attributes evaluation Purchase intention	All the analyzed parameters had an acceptance index above 75%. Flavour was the attribute with the lowest evaluation 63% of participants had a positive response regarding their intention to buy the product.
Schouteten et al. (2016)	Belgium	Mealworm (not specified)	Meat burgers, plant-based burgers and mealworm-based burgers	Three different conditions: Blind, expected and informed. Overall liking Emotional and sensory profile using the EmoSensory wheel questionnaire format (using a 5-point RATA scale)	Regardless of condition, the insect-based burger had the worst overall liking. Under the expected or informed condition overall liking of the insect-based burger was higher than with blind condition. Positive emotions were more frequent in the expected condition, but several negative emotions (distrust, fear or worried) were less frequent after tasting (blind and informed condition). Mealworm-based burger had a similar emotional profile to the plant-based burger but was associated with more negative and less positive attributes than the meat burger. The mealworm-based burger was associated with different attributes than the meat and plant-based burger. Those attributes were related to texture (more dry and granular and less juicy and soft), flavour (less meat flavour and higher nutty and off-flavour) and aroma (less meat aroma).

Reference	Country	Insect species	Product	Measures	Results
Tan et al. (2017b)	Netherlands	Mealworm *Tenebrio molitor*	Meatballs (with minced meat or with minced meat with ground mealworms) and sweet strawberry orange dairy drink with suspended cereal bits (without mealworms or with 5% ground mealworms).	Familiarity with tested food products Expected sensory liking Expected sensory profile (with check-all-that-apply question) Sensory liking Sensory profile (with check-all-that-apply question) Idea sensory profile Willingness to buy once or regularly mealworm products Appropriateness of food products	Acceptance was correlated with product appropriateness (higher for the meatballs), which disappeared after tasting. Both the acceptance and sensory liking was higher for the products not incorporating mealworms. After tasting both meatballs had lower liking, while the inverse occurred for the dairy drink. Willingness to buy the mealworm products decreased after tasting them. Addition of mealworms to the food products resulted in the association with less desirable sensory attributes. Meatballs with mealworms were more 'musty', 'mealy', 'nutty', 'dry', less 'meaty', 'fatty' and 'juicy' and had a longer 'aftertaste' than the control product. – Dairy drink with mealworms was more 'musty', 'mealy', 'rough', 'grainy' and less 'sweet' and 'thick' than the control product.
Caparros Megido et al. (2016)	Belgium	Mealworms *Tenebrio molitor*	Beef burger, beef/mealworm burger, lentil burger, beef/lentil burger	Overall liking Hedonic attributes evaluation	Burgers that included mealworm and/or lentils had poorer hedonic evaluations than the beef burger. The perception of insect consumption evolved positively in 84% of participants after the tasting session.
Caparros Megido et al. (2014)	Belgium	Mealworm *Tenebrio molitor* and cricket *Acheta domesticus*	Baked crickets, boiled crickets, baked mealworms, boiled mealworms, crushed mix of baked crickets and mealworms, baked mealworms with a pinch of vanilla or paprika, baked mealworms dunked in chocolate	Overall liking	Baked mealworms (natural), baked mealworms with paprika and baked mealworms dunked in chocolate had the highest scores of overall liking. Boiled crickets and mealworms had the poorest evaluations.

Europe and the U.S.A (Farina 2017; Caparros Megido et al. 2014, 2016; Ribeiro et al. 2018b; Schouteten et al. 2016; Sogari et al. 2018; Tan et al. 2017b; Tao et al. 2017; Zhong 2017; Zielińska et al. 2018) crickets, locusts and mealworms have been used, while those performed in Nigeria (Ogunlakin et al. 2018) and Brazil (de Oliveira et al. 2017) evaluated termites and cockroaches, respectively. While there are studies that evaluated whole/visible edible insects (Caparros Megido et al. 2014; Zielińska et al. 2018), the majority of the sensory evaluation studies were developed with insect-based food products including non-visible insects: cereal bars (Ribeiro et al. 2018b; Zhong 2017), burgers (Caparros Megido et al. 2016; Schouteten et al. 2016), biscuits (Ogunlakin et al. 2018), jelly (Sogari et al. 2018), rice (Tao et al. 2017), broth (Farina 2017), bread (de Oliveira et al. 2017), meatball and dairy drinks (Tan et al. 2017b). Although the appropriateness of the food product may influence the expected sensory-liking, this effect disappears after tasting the product (Tan et al. 2017b). Nonetheless, food appropriateness can have a great effect on future willingness to eat the products, especially when considering regular consumption (Tan et al. 2016a, 2017a, b).The use of processed insects (Zielińska et al. 2018) and their association with familiar flavours (Caparros Megido et al. 2014) increases their liking. However, Sogari et al. (2018) have also reported that a jelly incorporating whole crickets yielded higher liking scores and a more positive sensory profile than such jelly with cricket flour. The way insects are slaughtered can also have some effect on their sensory properties as reported by Farina (2017), where a broth made with frozen crickets had higher overall liking scores and better umami flavour than the same broth made with the addition of live crickets. The work developed by Ribeiro et al. (2018b), reports the impact of using different species of crickets in the production of a cereal bar, with the inclusion of *A. domesticus* leading to larger overall liking scores and a better sensory profile than the inclusion of *G. sigillatus*. Similarly, Tao et al. (2017) compared the incorporation of different species, reporting that extruded rice with *A. domesticus* flour presented better attributes and overall liking evaluation than rice with *L. migratoria* flour. Moreover, in the study by Zielińska et al. (2018), the sensory evaluation of whole insects was also accessed, comparing *S. gregaria*, *G. sigillatus* and *T. molitor*, with the first yielding a better sensory evaluation. Further studies should be developed to better grasp the impact of the use of different species and of different forms of incorporation (whole or processed) on both the overall liking and sensory profile of food products.

The comparison of the formulation of insect-based food products with their insect free counterparts, indicates that the inclusion of insects leads to lower hedonic scores, less willingness to eat and poorer sensory profiles associated with negative attributes (Caparros Megido et al. 2016; Ogunlakin et al. 2018; Ribeiro et al. 2018b; Schouteten et al. 2016; Tan et al. 2017b; Zhong 2017). It is acceptable to think that such differences in sensory evaluations are directly related to the sensory attributes of the edible insects, considering that the simple disclosure of the inclusion of insects in the food product is not enough to provoke significant changes in their liking or sensory profile (Tan et al. 2016a, 2017a). Despite being performed with different products: cereal bars (Ribeiro et al. 2018b), burgers (Schouteten et al. 2016), meatballs and dairy drink (Tan et al. 2017b), these studies have some similarities in the sensory profiles of the products: dry, mealy and grainy texture, unpleasant flavour and odour, existence of off-flavours and prolonged aftertaste. Furthermore, some of the attributes related to odour and flavour that were associated with the products incorporating insects (earthy, rancid, mouldy, musty) can be indicative of lipid oxidation (Paradiso et al. 2009). It is noteworthy that in the work developed by Ribeiro et al. (2018b), they reported that the defatting of crickets (from two different species) eliminated the association with negative attributes and significantly improved their overall liking and acceptability, bringing it to the same level as its insect free counterpart. This enforces the perception that the lipid fraction is responsible for the poor sensory evaluation.

Beyond sensory profiling, Schouteten et al. (2016) also assessed the emotions evoked by consuming edible insects. In their work, when participants tasted the insect-based food, the association with some negative emotions (worried, fear) but also with some positive emotions (glad, merry) has decreased, compared with the evaluation of the simple concept of eating the insect. When compared to a beef burger, the mealworm burger was associated with more negative emotions and less positive emotions, regardless of the conditions in which the tasting occurred. Furthermore, informing consumers about the practice of entomophagy decreased the association with 'distrust' when compared to tasting the burger containing insects without any prior information. Nonetheless, informing consumers about entomophagy only lead to slight improvements in the sensory profile and liking scores.

Conclusion

Entomophagy has a low acceptance rate in Western countries mainly due to food disgust and food neophobia, which are mostly caused by the lack of familiarity with edible insects and the association with lack of hygiene and dirtiness. In order to reverse this situation, different strategies have to be implemented, such as popularizing the concept of entomophagy and its advantages or incorporating insects into familiar dishes in a processed, visible/non-visible form in accordance with the characteristics of the food product. Moreover, there are still several issues affecting the way in which entomophagy is perceived in the West, as products have to be considered appropriate by consumers and should satisfy other factors (e.g., price, availability) to compete with similar food products already present in the market. The sensory properties of the foods containing insects should also be further explored as many lack sensory appeal. Such improvements are needed to drive consumers into the incorporation of edible insects their diets.

Bibliography

Al-Shawaf L, Lewis DMG, Alley TR, Buss DM (2015) Mating strategy, disgust, and food neopho-
bia. Appetite 85:30–35. https://doi.org/10.1016/j.appet.2014.10.029
Belluco S, Halloran A, Ricci A (2017) New protein sources and food legislation: the case of edible
insects and EU law. Food Security 9(4):803–814. https://doi.org/10.1007/s12571-017-0704-0
Caparros Megido R, Sablon L, Geuens M, Brostaux Y, Alabi T, Blecker C et al (2014) Edible
insects acceptance by Belgian consumers: promising attitude for entomophagy development.
J Sens Stud 29(1):14–20. https://doi.org/10.1111/joss.12077
Caparros Megido R, Gierts C, Blecker C, Brostaux Y, Haubruge É, Alabi T, Francis F (2016)
Consumer acceptance of insect-based alternative meat products in Western countries. Food
Qual Prefer 52:237–243. https://doi.org/10.1016/j.foodqual.2016.05.004
Castro M, Chambers E IV (2018) Willingness to eat an insect based product and impact on brand
equity: a global perspective. J Sens Stud 0(0):e12486. https://doi.org/10.1111/joss.12486
Chan EY (2019) Mindfulness and willingness to try insects as food: the role of disgust. Food Qual
Prefer 71:375–383. https://doi.org/10.1016/j.foodqual.2018.08.014
Chapman HA, Anderson AK (2012) Understanding disgust. Ann N Y Acad Sci 1251:62–76.
https://doi.org/10.1111/j.1749-6632.2011.06369.x
Clarkson C, Mirosa M, Birch J (2018) Consumer acceptance of insects and ideal product attri-
butes. Br Food J 120(12):2898–2911. https://doi.org/10.1108/BFJ-11-2017-0645
Cunha LM, Moura AP, Costa-Lima R (2014) Consumers' associations with insects in the context
of food consumption: comparisons from acceptors to disgusted. Oral presentation at the insects
to feed the world, 14–17 May, The Netherlands
Cunha LM, Gonçalves ATS, Varela P, Hersleth M, Costa Neto EM, Grabowski NT, et al (2015)
Adoption of insects as a source for food and feed production: a cross-cultural study on determinants of acceptance. Oral presentation at the "11th Pangborn sensory science symposium",
23–27 August, Gothenburg, Sweden
Cunha LM, Cabral D, Moura AP, de Almeida MDV (2018) Application of the food choice questionnaire across cultures: systematic review of cross-cultural and single country studies. Food
Qual Prefer 64:21–36. https://doi.org/10.1016/j.foodqual.2017.10.007
de Oliveira LM, da Silva Lucas AJ, Cadaval CL, Mellado MS (2017) Bread enriched with flour
from cinereous cockroach (Nauphoeta cinerea). Innovative Food Sci Emerg Technol 44:30–35.
https://doi.org/10.1016/j.ifset.2017.08.015
De Rosa B, Franco S, Lacetera N, Cicatiello C (2016) Consumer approach to insects as food:
barriers and potential for consumption in Italy. Br Food J 118(9):2271–2286. https://doi.
org/10.1108/BFJ-01-2016-0015
de-Magistris T, Pascucci S, Mitsopoulos D (2015) Paying to see a bug on my food: how regulations and information can hamper radical innovations in the European Union. Br Food
J 117(6):1777–1792. https://doi.org/10.1108/BFJ-06-2014-0222
Deroy O, Reade B, Spence C (2015) The insectivore's dilemma, and how to take the West out of it.
Food Qual Prefer 44:44–55. https://doi.org/10.1016/j.foodqual.2015.02.007
Farina MF (2017) How method of killing crickets impact the sensory qualities and physiochemical
properties when prepared in a broth. Int J Gastron Food Sci 8:19–23. https://doi.org/10.1016/j.
ijgfs.2017.02.002
Fischer ARH, Steenbekkers LPAB (2018) All insects are equal, but some insects are more equal
than others. Br Food J 120(4):852–863. https://doi.org/10.1108/BFJ-05-2017-0267
Gmuer A, Guth J, Hartmann C, Siegrist M (2016) Effects of the degree of processing of insect
ingredients in snacks on expected emotional experiences and willingness to eat. Food Qual
Prefer 54:117–127. https://doi.org/10.1016/j.foodqual.2016.07.003
Haidt J, McCauley C, Rozin P (1994) Individual differences in sensitivity to disgust: a scale sampling seven domains of disgust elicitors. Personal Individ Differ 16(5):701–713. https://doi.
org/10.1016/0191-8869(94)90212-7

Hamerman EJ (2016) Cooking and disgust sensitivity influence preference for attending insect-based food events. Appetite 96:319–326. https://doi.org/10.1016/j.appet.2015.09.029

Hartmann C, Siegrist M (2016) Becoming an insectivore: results of an experiment. Food Qual Prefer 51:118–122. https://doi.org/10.1016/j.foodqual.2016.03.003

Hartmann C, Siegrist M (2018) Development and validation of the food disgust scale. Food Qual Prefer 63:38–50. https://doi.org/10.1016/j.foodqual.2017.07.013

Hartmann C, Shi J, Giusto A, Siegrist M (2015) The psychology of eating insects: a cross-cultural comparison between Germany and China. Food Qual Prefer 44:148–156. https://doi.org/10.1016/j.foodqual.2015.04.013

Hartmann C, Ruby MB, Schmidt P, Siegrist M (2018) Brave, health-conscious, and environmentally friendly: positive impressions of insect food product consumers. Food Qual Prefer 68:64–71. https://doi.org/10.1016/j.foodqual.2018.02.001

House J (2016) Consumer acceptance of insect-based foods in the Netherlands: academic and commercial implications. Appetite 107(Supplement C):47–58. https://doi.org/10.1016/j.appet.2016.07.023

House J (2018) Insects are not 'the new sushi': theories of practice and the acceptance of novel foods AU – House. Jonas Social Cultural Geograp:1–22. https://doi.org/10.1080/14649365.2018.1440320

Jensen NH, Lieberoth A (2019) We will eat disgusting foods together – evidence of the normative basis of Western entomophagy-disgust from an insect tasting. Food Qual Prefer 72:109–115. https://doi.org/10.1016/j.foodqual.2018.08.012

La Barbera F, Verneau F, Amato M, Grunert K (2018) Understanding Westerners' disgust for the eating of insects: the role of food neophobia and implicit associations. Food Qual Prefer 64:120–125. https://doi.org/10.1016/j.foodqual.2017.10.002

Looy H, Dunkel FV, Wood JR (2014) How then shall we eat? Insect-eating attitudes and sustainable foodways. Agric Hum Values 31(1):131–141. https://doi.org/10.1007/s10460-013-9450-x

Martins Y, Pliner P (2006) "Ugh! That's disgusting!": identification of the characteristics of foods underlying rejections based on disgust. Appetite 46(1):75–85. https://doi.org/10.1016/j.appet.2005.09.001

Menozzi D, Sogari G, Veneziani M, Simoni E, Mora C (2017a) Eating novel foods: an application of the theory of planned behaviour to predict the consumption of an insect-based product. Food Qual Prefer 59:27–34. https://doi.org/10.1016/j.foodqual.2017.02.001

Menozzi D, Sogari G, Veneziani M, Simoni E, Mora C (2017b) Explaining the intention to consume an insect-based product: a cross-cultural comparison. In: Seyal AH, Rahman MNA (eds) Theory of planned behavior: new research. Nova Publisher, New York

Myers G, Pettigrew S (2018) A qualitative exploration of the factors underlying seniors' receptiveness to entomophagy. Food Res Int 103:163–169. https://doi.org/10.1016/j.foodres.2017.10.032

Nonaka K (2009) Feasting on insects. Entomol Res 39(5):304–312. https://doi.org/10.1111/j.1748-5967.2009.00240.x

Ogunlakin GO, Oni VT, Olaniyan SA (2018) Quality evaluation of biscuit fortified with edible termite (Macrotermes nigeriensis). Asian J Biotechnol Bioresour Technol 4(2). https://doi.org/10.9734/AJB2T/2018/43659

Paradiso VM, Summo C, Pasqualone A, Caponio F (2009) Evaluation of different natural antioxidants as affecting volatile lipid oxidation products related to off-flavours in corn flakes. Food Chem 113(2):543–549. https://doi.org/10.1016/j.foodchem.2008.07.099

Pelchat ML, Pliner P (1995) "Try it. You'll like it". Effects of information on willingness to try novel foods. Appetite 24(2):153–165. https://doi.org/10.1016/S0195-6663(95)99373-8

Randrianandrasana M, Berenbaum MR (2015) Edible non-crustacean arthropods in rural communities of Madagascar. J Ethnobiol 35(2):354–383. https://doi.org/10.2993/etbi-35-02-354-383.1

Reinbold K, Pecoraro NM, Frøst MB (2018) Developing novel foods with insects – to see or not to see. Poster presentation at the EuroSense'18 – 8th European conference on sensory and consumer research, Verona, Italy

Ribeiro JC, Cunha LM, Sousa-Pinto B, Fonseca J (2018a) Allergic risks of consuming edible insects: a systematic review. Mol Nutr Food Res. https://doi.org/10.1002/mnfr.201700030

Ribeiro JC, Cunha LM, Lima RC, Maia MRG, Cabrita AR (2018b) Increasing liking and improving sensory profile of cereal bars incorporating dried edible crickets: impact of defatting. Oral presentation at the EuroSense'18 – 8th European conference on sensory and consumer research, Verona, Italy

Rozin P, Haidt J (2013) The domains of disgust and their origins: contrasting biological and cultural evolutionary accounts. Trends Cogn Sci 17(8):367–368. https://doi.org/10.1016/j.tics.2013.06.001

Rozin P, Millman L, Nerneroff C (1986) Operation of the laws of sympathetic magic in disgust and other domains. J Pers Soc Psychol 50(4):703–712. https://doi.org/10.1037/0022-3514.50.4.703

Ruby MB, Rozin P, Chan C (2015) Determinants of willingness to eat insects in the USA and India. J Insects Food Feed 1(3):215–225. https://doi.org/10.3920/JIFF2015.0029

Rumpold BA, Schluter OK (2013) Nutritional composition and safety aspects of edible insects. Mol Nutr Food Res 57(5):802–823. https://doi.org/10.1002/mnfr.201200735

Schouteten JJ, De Steur H, De Pelsmaeker S, Lagast S, Juvinal JG, De Bourdeaudhuij I et al (2016) Emotional and sensory profiling of insect-, plant- and meat-based burgers under blind, expected and informed conditions. Food Qual Prefer 52:27–31. https://doi.org/10.1016/j.foodqual.2016.03.011

Sogari G, Menozzi D, Mora C (2017) Exploring young foodies' knowledge and attitude regarding entomophagy: a qualitative study in Italy. Int J Gastron Food Sci 7:16–19. https://doi.org/10.1016/j.ijgfs.2016.12.002

Sogari G, Menozzi D, Mora C (2018) Sensory-liking expectations and perceptions of processed and unprocessed insect products. Int J Food Syst Dyn 9(4):314–320. https://doi.org/10.18461/ijfsd.v9i4.942

Sogari G, Menozzi D, Mora C (2019) The food neophobia scale and young adults' intention to eat insect products. Int J Consum Stud 43(1):68–76. https://doi.org/10.1111/ijcs.12485

Tan HSG, Fischer ARH, Tinchan P, Stieger M, Steenbekkers LPA, van Trijp HCM (2015) Insects as food: exploring cultural exposure and individual experience as determinants of acceptance. Food Qual Prefer 42(Supplement C):78–89. https://doi.org/10.1016/j.foodqual.2015.01.013

Tan HSG, Fischer ARH, van Trijp HCM, Stieger M (2016a) Tasty but nasty? Exploring the role of sensory-liking and food appropriateness in the willingness to eat unusual novel foods like insects. Food Qual Prefer 48:293–302. https://doi.org/10.1016/j.foodqual.2015.11.001

Tan HSG, Berg E v d, Stieger M (2016b) The influence of product preparation, familiarity and individual traits on the consumer acceptance of insects as food. Food Qual Prefer 52:222–231. https://doi.org/10.1016/j.foodqual.2016.05.003

Tan HSG, Tibboel CJ, Stieger M (2017a) Why do unusual novel foods like insects lack sensory appeal? Investigating the underlying sensory perceptions. Food Qual Prefer 60:48–58. https://doi.org/10.1016/j.foodqual.2017.03.012

Tan HSG, Verbaan YT, Stieger M (2017b) How will better products improve the sensory-liking and willingness to buy insect-based foods? Food Res Int 92:95–105. https://doi.org/10.1016/j.foodres.2016.12.021

Tao J, Davidov-Pardo G, Burns-Whitmore B, Cullen EM, Li YO (2017) Effects of edible insect ingredients on the physicochemical and sensory properties of extruded rice products. J Insects Food Feed 3(4):263–278. https://doi.org/10.3920/JIFF2017.0030

van Huis A, Van Itterbeeck J, Klunder HC, Mertens E, Halloran A, Muir G, Vantomme P (2013) Edible insects: future prospects for food and feed security. FAO, Rome

Vandeweyer D, Crauwels S, Lievens B, Van Campenhout L (2017) Microbial counts of mealworm larvae (Tenebrio molitor) and crickets (Acheta domesticus and Gryllodes sigillatus) from different rearing companies and different production batches. Int J Food Microbiol 242:13–18. https://doi.org/10.1016/j.ijfoodmicro.2016.11.007

Verbeke W (2015) Profiling consumers who are ready to adopt insects as a meat substitute in a Western society. Food Qual Prefer 39:147–155. https://doi.org/10.1016/j.foodqual.2014.07.008

Verneau F, La Barbera F, Kolle S, Amato M, Del Giudice T, Grunert K (2016) The effect of communication and implicit associations on consuming insects: an experiment in Denmark and Italy. Appetite 106:30–36. https://doi.org/10.1016/j.appet.2016.02.006

Woolf E, Zhu Y, Emory K, Zhao J, Liu C (2019) Willingness to consume insect-containing foods: a survey in the United States. LWT Food Sci Technol 102:100–105. https://doi.org/10.1016/j.lwt.2018.12.010

Yen AL (2009) Edible insects: traditional knowledge or Western phobia? Entomol Res 39(5):289–298. https://doi.org/10.1111/j.1748-5967.2009.00239.x

Zhong A (2017) Product development considerations for a nutrient rich bar using cricket (Acheta domesticus) protein. (Master of Science in Nutritional Science), California State University, ProQuest

Zielińska E, Karaś M, Baraniak B (2018) Comparison of functional properties of edible insects and protein preparations thereof. LWT Food Sci Technol 91:168–174. https://doi.org/10.1016/j.lwt.2018.01.058

Chapter 6
Quality and Consumer Acceptance of Products from Insect-Fed Animals

Laura Gasco, Ilaria Biasato, Sihem Dabbou, Achille Schiavone, and Francesco Gai

Abstract Fish and soybean meal are the most common protein sources in aquaculture and poultry feed ingredients, but these conventional sources are no longer sustainable and will be further limited by increasing prices. New and sustainable protein sources for animal feeds are necessary, and insects seem a promising, novel option due to their good nutritional profile and lower environmental impact. After a brief introduction, this chapter critically reviews the latest knowledge about the dietary use of insect meals in fish, shellfish and avian species. Particular focus is put on their impact on the flesh and meat of aquaculture and poultry products in terms of sensorial perception and quality traits. In general, analysis of sensory properties shows that for both products no differences were perceived if untrained panelists were involved in the sensorial analysis. Concerning meat and flesh quality, results are controversial, but a dramatic influence of insect meal fatty acid (FA) profile with a decrease in long chain n-3 FA content has been observed in both species. Moreover, an overview on the available data about consumer acceptance towards food products from insects-fed animals is provided.

Keywords Aquaculture · Alternative proteins · Animal origin food · Sensory · Egg · Insects · Meat · Poultry

L. Gasco (✉) · I. Biasato
University of Turin, Department of Agricultural, Forest, and Food Sciences, Grugliasco, Italy
e-mail: laura.gasco@unito.it; ilaria.biasato@unito.it

S. Dabbou · A. Schiavone
University of Turin, Department of Veterinary Sciences, Grugliasco, Italy
e-mail: sihem.dabbou@unito.it; achille.schiavone@unito.it

F. Gai
Italian National Research Council, Institute of Food Production Sciences, Grugliasco, Italy
e-mail: francesco.gai@ispa.cnr.it

© Springer Nature Switzerland AG 2019
G. Sogari et al. (eds.), *Edible Insects in the Food Sector*,
https://doi.org/10.1007/978-3-030-22522-3_6

Abbreviations

DHA   Docosahexaenoic acid
EPA   Eicosapentaenoic acid
EU   European Union
FA   Fatty Acid
FM   Fish Meal
HI   Hermetia Illucens
IM   Insect Meal
MD   Musca Domestica
PAP   Processed Animal Proteins
SBM   Soybean Meal
SFA   Saturated Fatty Acids
TM   Tenebrio Molitor

Introduction

Fish meal (FM) and soybean meal (SBM) are the most common protein sources in aquaculture and poultry feed, but these conventional sources are no longer sustainable and will be further limited by increasing prices (Veldkamp and Bosch 2015). New and sustainable protein sources for animal feeds are necessary and insects seem to be promising alternatives due to their good nutritional profile and low environmental impact (Van Huis and Oonincx 2017).

The European Union (EU) Commission recently approved the use of Processed Animal Proteins (PAPs) from seven insect species for aquaculture feeds (Regulation (EU) 2017/893). Moreover, the EU Commission amended Regulation (EU) 68/2013 on the Catalogue of feed materials, introducing "terrestrial live invertebrates" or "dead invertebrates with or without treatment but not processed" as referred to in Regulation (EC) 1069/2009 (Regulation (EU) 2017/1017). Of course, these land animal products shall fulfil the requirements of the Regulation (EC) 1069/2009 and Regulation (EU) 142/2011 and may be subject to restrictions in use according to Regulation (EC) 999/2001.

Thus, under the current EU Regulations, PAPs from insects can only be used for aquaculture, while live or not processed dead insects can also be used as feed in in monogastric animals.

Outside of the EU, different regulations exist and other insect species may be used for feed purposes. Overall, there is great interest in using insects (raw or processed) for animal feed (Biasato et al. 2017, 2018; Dobermann et al. 2017; Henry et al. 2015; Józefiak et al. 2016; Makkar et al. 2014; Sánchez-Muros et al. 2014). The sections that follow provide an evaluation of the dietary use of insect meals (IM) in fish, shellfish and avian species, specifically their impact on sensorial perception and quality traits of the flesh and meat of aquaculture and poultry products.

Aquaculture Products

Due to the increasing interest in the use of IM in aquafeeds, a consistent number of nutritional studies have been carried out in both fish and shellfish species; their impact on product quality is reported in Table 6.1.

The effects of dietary IM inclusion on proximate composition and quality parameters of aquaculture products were investigated testing different inclusion levels of *Hermetia illucens* (HI) and *Tenebrio molitor* (TM) meals. Proximate composition of shrimp muscle and fillets of rainbow trout (raw and cooked) fed TM diets were evaluated; no differences were recorded in measures of moisture, protein and ash content (Panini et al. 2017; Iaconisi et al. 2018). Contrastingly, an increase in dry matter and ether extract contents of trout dorsal fillets was found by Renna et al. (2017) in fish fed the highest (50%) dietary HI larvae meal inclusion.

Concerning the quality parameters, studies carried out on blackspot seabream, gilthead seabream and rainbow trout fed with different TM meal inclusion levels reported no significant differences in some fillet quality parameters, such as water holding capacity and texture characteristics (Iaconisi et al. 2017, 2018; Piccolo et al. 2017). As far as the fish colour is concerned, TM diets may affect the colours of the fillet and skin of blackspot seabream. In particular, the highest redness index (a*) in the skin ventral region and an increased yellowness (b*) in the fillet epaxial region were found in fish-fed the maximum inclusion level of TM (Iaconisi et al. 2017). Conversely, results reported by Mancini et al. (2018) highlighted a decreased fillet yellowness in rainbow trout fed HI diet replacing 50% of FM. The authors explained this opposite trend with a modification of the fatty acid (FA) muscle profile related to a different FA profile of the IM utilised in the trials.

Since IM use in fish diets may lead to changes in fillet fatty acid composition, sensory properties of fish products can vary as well (in particular the aroma and flavour, which are directly linked to the dietary lipid-volatile components) (Turchini et al. 2007; Borgogno et al. 2017). In general, capability of perceiving sensory differences may depend on training of panellists. For instance, unaffected sensory parameters were perceived by both untrained and trained panellists in rainbow trout (Sealey et al. 2011) and Atlantic salmon (Lock et al. 2016) fed with HI prepupae and larvae meals, respectively, as partial or total replacement of FM. However, a more recent rainbow trout feeding trial involving untrained panellists did not highlight any significant differences for any selected parameters of taste and odour, while a significantly darker filet colour was identified in fish fed with HI meal compared to a control diet (Stadtlander et al. 2017). Contrastingly, significant changes in perceived intensity of aroma, flavour and texture descriptors of rainbow trout fed with HI meal as FM replacer were highlighted by Borgogno et al. (2017) who used trained panellists. Specifically, the dominance of metallic flavour characterised fillets of fish fed HI diets, demonstrating an unfamiliar flavour to the consumer. Nevertheless, the authors concluded that dietary IM inclusion did not induce the perception of off-flavours.

Table 6.1 Maximum level of FM substitution (and IM inclusion) and related impacts on flesh quality traits

Fish /Shellfish Species		Insect species tested	Max % of FM substitution	% IM inclusion	Major impacts on product quality	Reference
Atlantic salmon	*Salmo salar*	HI	100	5–10–25	Sensory testing of fillets of fish fed 10 and 25 IM inclusion level did not reveal any significant differences in odour, flavour/taste or texture between groups	Lock et al. (2016)
Carp var. Jian	*Cyprinus carpio*	HI	100	3.5–14	No differences in proximate composition while HI inclusion decrease the n-3 highly unsaturated fatty acid composition in body of fish.	Zhou et al. (2018)
Rainbow trout	*Oncorhynchus mykiss*	HI	50	25–50	Significant changes in perceived intensity of aroma, flavor and texture. Dominance of metallic flavor characterized fillets of fish fed HI diets.	Borgogno et al. (2017) Renna et al. (2017) Mancini et al. (2018) Secci et al. (2018a)
Rainbow trout	*Oncorhynchus mykiss*	HI	50	50	No differences except a slightly darker coloration of fish fed HI were observed in a controlled panel test.	Stadtlander et al. (2017)
Rainbow trout	*Oncorhynchus mykiss*	HI	50	25–50	No significant difference were observed in a controlled panel test of fish fed the FM containing control diet as compared to fish fed the enriched HI or HI diets.	Sealey et al. (2011)

(continued)

Table 6.1 (continued)

Fish /Shellfish Species	Insect species tested	Max % of FM substitution	% IM inclusion	Major impacts on product quality	Reference
Rainbow trout	*Oncorhynchus mykiss*	TM	67	25–50	No negative effect on most quality traits of the fish flesh. The fatty acids C16:0, C18:1n9 and C18:2n6 increased whilst EPA and DHA progressively diminished in fillets when TM inclusion in feeds increased	Belforti et al. (2015) Iaconisi et al. (2018)
Gilthead seabream	*Sparus aurata*	TM	74	25–50	No negative effect on marketable indexes with a 25% of TM inclusion level. At 50% of TM inclusion level dressed yield was penalized.	Piccolo et al. (2017)
Blackspot seabream	*Pagellus bogaraveo*	TM	50	25–50	TM dietary inclusion affect some fillet quality parameters as ventral colour and muscle fatty acid profile.	Iaconisi et al. (2017)
Pacific white shrimp	*Litopenaeus vannamei*	TM	100	7.6–30.5	Colour and firmness were unchanged between the treatments. Dietary TM affected the lipid and fatty acid composition of shrimp muscle.	Panini et al. (2017)
Pacific white shrimp	*Litopenaeus vannamei*	TM	100	7.0–36	Maximum whole-body protein and lipid content achieved when HI inclusion was restricted to 29% and 15%, respectively.	Cummins Jr et al. (2017)

DHA Docosahexaenoic acid, *EPA* Eicosapentaenoic acid, *FM* fish meal, *HI Hermetia illucens*, *TM Tenebrio molitor*

In terms of lipid profile, insect larvae are characterized by poor contents of highly polyunsaturated fatty acids (PUFA). In fact, in land-based products (including SBM) and insects, the long chain FA (eicosapentaenoic acid, EPA and docosahexaenoic acid, DHA) are usually absent. Insect FA profiles may greatly vary with the insect species and substrates used for their rearing (Gasco et al. 2018), thus also affecting the fish products. Due to the high content of saturated fatty acids (SFA) of HI, freshwater fish fed with increasing levels of HI meal showed increased contents of SFA (mostly lauric acid, C12:0) and decreased contents of valuable PUFA (both n-3 and n-6) (Renna et al. 2017; Mancini et al. 2018; Secci et al. 2018a; Zhou et al. 2018). Contrastingly, TM is characterised by high contents of oleic, linoleic and palmitic acids (Gasco et al. 2018). Fish fed diets including high levels of TM meal showed increased n-6 PUFA contents at the expense of n-3 polyunsaturated content (Belforti et al. 2015; Iaconisi et al. 2017, 2018), with a consequent reduction of the Σn-3/Σn-6 FA ratio and a worsening of the atherogenicity and thrombogenicity indexes.

Compared to other aquaculture products such as shellfish (in particular shrimp), dietary IM inclusion and their effects on product quality are poorly investigated. So far only a couple of papers have investigated the use of TM and HI in diets for Pacific white shrimp (*Litopenaeus vannamei*). Panini et al. (2017) concluded that dietary TM meal inclusion did not affect the Pacific white shrimp muscle quality, even if inclusion levels above the 25% FM substitution showed increased lipid and decreased PUFA muscle contents. Contrastingly, Cummins et al. (2017) tested different inclusion levels of HI meals and showed that the maximum whole-body lipid content could be achieved with a 15% of inclusion, given, however, no information about the FA profile of these products.

Poultry Products

Despite increasing interest in the use of IM in poultry feeds (in addition to fish feeds), a limited number of studies assessing products quality has been carried out until now. The current scientific research that has highlighted the impact of IM use on carcass characteristics and meat quality products are reported in Table 6.2.

Concerning meat quality, results are controversial. Cullere et al. (2016) observed that the redness index in the breast meat of broiler quails was affected by increasing dietary inclusion levels of HI larva meal. However, a partial substitution (25% or 50%) of dietary soybean protein with TM and HI meals in Barbary partridges (*Alectoris barbara*) has been reported to not affect the pH and colour of the raw meat, even if the presence of IM seemed to increase the yellowness index of the cooked meat (Secci et al. 2018b). Contrastingly, Altmann et al. (2018), Pieterse et al. (2018) and Leiber et al. (2017) did not find any significant effects of dietary HI meal inclusion on broiler meat colour. The inclusion of MD larva meal in broiler diets has also been associated with a significant decrease in breast muscle lightness (Pieterse et al. 2014). However, Bovera et al. (2016) did

Table 6.2 Maximum level of SBM substitution (and IM inclusion) and related impacts on egg and meat quality traits

Avian species	Insect species	Max % of SBM/FM substituted	% IM inclusion	Days of feeding	Major impacts on product quality	Reference
Barbary partridges	HI TM	68 SBM	12–22 TM 10–19 HI	64	No differences in the whole body composition except for the ash content.	Secci et al. (2018b)
					The carcass weights of all the insect groups were higher than the SBM group.	Loponte et al. (2017)
Broiler chickens	BM MD TM	100 SBM	7.8 BM 8.0 MD 8.1 TM	35	Tenderness and juiciness of meat were higher in TM group compared to the control and other treatments.	Khan et al. (2018)
Broiler chickens	HI	50 SBM	11.9–14.5	34	HI meal results in a product that does not differ from the standard fed control group, with the exception that the breast filet has a more intense flavour that decreases over storage time.	Altmann et al. (2018)
		49 SBM	7.8	75	Regarding quality parameters, only cooking loss was increased with the HI plus pea protein diet compared with the control.	Leiber et al. (2017)
		64 SBM	5–10–15	49	Replacement of SBM and FM with HI meal did not affect aroma or taste of cooked breast meat.	Onsongo et al. (2018)
		Not specified	5–10–15	32	No significant differences for pH, colour, thaw loss and cooking loss as well on the sensory characteristics (aroma, flavour, juiciness and tenderness) of the breast muscle of the broilers fed HI meal.	Pieterse et al. (2018)
Broiler chickens	MD	100 FM	10	32	Meat quality parameters were not affected except for drip loss that were lowest in HI meal treated group.	Pieterse et al. (2014)
					Higher sustained juiciness values was found in chicken larvae fed.	

(continued)

Table 6.2 (continued)

Avian species	Insect species	Max % of SBM/FM substituted	% IM inclusion	Days of feeding	Major impacts on product quality	Reference
Laying hens	HI	100 SBM	17	147	Hens fed the insect-based diet(HIM) produced eggs with a higher proportion of yolk than the group fed the SBM group. HIM was associated with redder yolks, richer in γ-tocopherol, lutein, β-carotene and total carotenoids than SBM yolks.	Secci et al. (2018c)
Laying hens	HI	41 SBM	5–7.5	182	Hens fed the HI based diet linearly increased yolk color, egg shell-breaking strength and egg thickness.	Mwaniki et al. (2018)
Laying hens	HI	39 SBM	3.5–5–6.5	112	Hens fed HI diet showed higher egg production, egg weight and values of Haugh unit and egg shell thickness compared to those of the control.	Park et al. (2017)
Quail	HI	24.8 SBM	10–15	28	Breast meat weight and yield did not differ while the inclusion of HI meal reduced meat pHu. Meat proximate composition, cholesterol content and oxidative status remained unaffected by HI supplementation as well as its sensory characteristics and off-flavours perception.	Cullere et al. (2016, 2018)

BM Bombyx mori, FM Fish meal, *HI Hermetia illucens, MD Musca domestica, SBM* soybean meal, *TM Tenebrio molitor*

not find any significant effects on the colour of raw and cooked meat, or on the skin of broiler chickens, also showing that consumers could accept the meat from broilers fed with TM meal.

Studies on the effects of dietary IM inclusion on poultry meat proximate composition also conflict. Cullere et al. (2018), Pieterse et al. (2018) and Secci et al. (2018b) did not report any significant effects on meat chemical composition of broiler quails, chickens or Barbary partridges fed diets with either HI or TM meals. Contrastingly, Ballitoc and Sun (2013) reported the highest percentage of breast fat content in broiler chickens fed with the highest level of dietary TM meal inclusion.

In terms of sensory characteristics of poultry products, research conducted in Nigeria showed that meat obtained from broilers fed MD diets did not reveal any distinctive organoleptic qualities, and was accepted by consumers (Awoniyi 2007). Cullere et al. (2018) and Onsongo et al. (2018) did not report any defects or off-flavour, nor aroma or taste problems, that could negatively influence the consumer acceptability of meat obtained from broiler quails or chickens fed different inclusion levels of HI. Similarly, Khan et al. (2018) reported that different IM products did not affect meat taste or flavour, but tenderness and juiciness were higher in the TM group compared to the control and other diets. Contrastingly, Altmann et al. (2018) showed that breast meat of broiler fed diets containing HI meals had a more intense flavour that decreased over storage time. Finally, Pieterse et al. (2014) found that the sensory profile of meat derived from chickens fed with diets containing MD larvae meal was slightly different from the control group because of a higher perception of metallic aroma and aftertaste, but a higher sustained juiciness and a lower mealiness in the mouth. The authors also reported that this specific aroma and aftertaste could potentially be attributed to the increased iron content of the larvae meal.

Like with fish, the use of IM in poultry feeds can dramatically influence the FA profile of poultry meat. Cullere et al. (2018) showed that dietary HI meal inclusion greatly affected the FA profile of Japanese quail breast meat. In particular, increasing levels of HI larvae meal lowered the healthiness of the meat as saturated fatty acids (SFA) increased at the expense of polyunsaturated fatty acids (PUFA). In a recent study by Secci et al. (2018b) about Barbary partridges fed diets containing IM in partial substitution of SMB, the HI and TM groups showed significantly higher oleic acid (C18:1n-9) and lower palmitic acid (C16:0) contents than the SBM group. The authors also highlighted that dietary HI meal inclusion induced a significant increase in lauric acid (C12:0) and palmitoleic acid (C16:1n-7) contents.

Concerning laying hens, Secci et al. (2018c) recently tested the effects of total replacement of SBM with HI larva meal in laying hens' diets (Lohmann Brown Classic) for 21 weeks, observing a higher proportion of yolk in the eggs, as well as higher amount of γ-tocopherol, lutein, β-carotene and total carotenoids, in the HI group. In another study, Mwaniki et al. (2018) reported that including up to 7.5% of defatted HI larvae meal in a corn–SBM diet for pullets (19 to 27 wk. of age) resulted in an increased yolk colour, egg shell-breaking strength and thickness. Hens fed diets containing HI have also been reported to show higher egg production and weight than those fed with a control diet (Park et al. 2017). Contrastingly, dietary HI larvae meal utilization in free range laying hens may result in a reduction of egg weight, shell weight and thickness, and yolk colour (Ruhnke et al. 2018). Finally, MD maggot meal has been reported to replace 50% of FM in diets for hens (5% of inclusion) without any adverse effects on egg production and shell strength (Agunbiade et al. 2007). However, the total replacement was deleterious to hen-egg production (Agunbiade et al. 2007).

Consumer Acceptance Towards Food Products from Insects-Fed Animals

Another important aspect in facing the impact of innovative feed ingredients in animal nutrition is represented by the consumer attitude towards novel food products. The determining factors in the buying process for several novel foods have been reported to mainly depend on the type of innovation and its market acceptance (Barrena and Sánchez 2012).

So far, data about consumer attitudes towards the utilization of insects in animal feeding are still lacking. In the first available survey involving 1300 consumers across 71 countries in the UK, EU and the Far East (East Asia, Russian Far East and Southeast Asia), the EU-funded PROteINSECT project showed that 73% of consumers were willing to eat fish, chickens or pork from animals fed on a diet containing insect protein. Furthermore, over 80% of people surveyed wanted to know more about insect utilization, with 64% recognizing no or low risk to human health in eating farmed animals fed insect meal (PROteINSECT 2016).

In a more recent survey involving 82 people, Verbeke et al. (2015) reported that 68% of the interviewed farmers, agriculture sector stakeholders, and citizens from Belgium were willing to accept the use of insects as feed ingredients in animal nutrition, especially for fish and poultry feed. The most relevant perceived benefits for the citizens were that the use of insects might allow a better exploitation of organic waste and lower dependence on foreign protein sources, as well as an improvement in the sustainability of livestock production and reduction of the ecological footprint of livestock (Verbeke et al. 2015).

In the same year, Neves (2015) recruited 363 and 303 Norwegian and Portuguese consumers, respectively, to test their acceptance of insects as feed. The obtained results revealed high acceptance to use insects to feed fish in both countries, with significantly higher acceptance among Norwegian consumers. A subsequent French survey conducted with 327 participants showed that the majority of consumers were willing to accept trout fed with insects when they have been informed of the environmental impact of the conventional feeding method and that the trout price was lower (Bazoche and Poret 2016).

The most recent European survey included a total of 4 stakeholders and 180 consumers from Scotland and was focused on the attitude towards the incorporation of cultured insect larvae- (maggots) derived feed materials into commercially formulated fish feeds for the Scottish salmon farming sector (Popoff et al. 2017). The results were promising for both survey categories. First, feed and salmon producers were generally open to the use of insect meals, provided the feeds were safe, reliable, and competitive and there were additional value benefits for producers. Second, the majority of consumers were also prepared to eat insect-fed fish with no concerns, while the 36% indicated specific conditions of unchanged price, safety and taste. It is also important to underline that most people favoured supermarket food and vegetable waste as rearing materials for insects, with only a minority considering animal manure, abattoir waste and human sewage suitable, thus also influencing their willingness to pay for the fish (Popoff et al. 2017).

Mancuso et al. (2016) recently explored the attitude and behaviour of Northern-Italian consumers of farmed fish fed with insects. Authors considered the different phases of the purchasing process, from interest in marine ecology and awareness of limited resources for fish feeding, to attitudes about eating finfish products if fed with insects, and finally to the decision to purchase. According to their findings, almost 90% of consumers were interested in research on more sustainable sources of feed used in aquaculture, also showing a positive attitude towards insect meals as feed in fish farming. In regards to purchasing activity, most of the respondents (76%) intended to buy and eat farmed fish even those fed insect meals, so long as hygiene requirements were met. About half of respondents (46.2%) also believed that the price would be the same as traditional fish products, whereas 29.2% and 23.8% thought that the product would have a lower or higher price, respectively, when compared to conventional (Mancuso et al. 2016). Another Italian survey evaluated the willingness of 341 consumers (students and employees from a university and ordinary citizens) to adopt insects as part of animal and human diets (Laureati et al. 2016). According to their findings, approximately 53% of the consumers appeared to be ready to incorporate insects into animal diets and to eat fish and livestock reared with insect-based feed. This outcome was attributed to the fact that fish and many other farmed animals (such as poultry and pigs) eat insects when they are reared in natural environments. Therefore, this phenomenon could have made the consumers more willing to accept the systematic use of insects or derivatives (e.g., meals) in farming. Interestingly, males were significantly more willing than females to consume products from insect-fed animals. Younger consumers, as well as people with a higher level of education about the topic (i.e., university students and employees) were also significantly more willing to accept insects as feed (Laureati et al. 2016).

Conclusions and Future Perspectives

In light of the considerations made in the previous sections, it is clear that in order to cope with an increasing global population and changing diets, an urgent supply of protein from sustainable sources for animal feeding is needed, especially in Europe where 70% of the protein is currently imported for animal feed purposes. Because of their good nutritional profile and lower environmental impact, the introduction of insects in the formulation of aquaculture and poultry feed ingredients should be considered as a beneficial long-term solution for sustainability and environmental impact. Available scientific literature demonstrates that from a technical point of view a partial or total replacement of conventional protein sources by means of insect proteins is feasible with minimal impact on the sensorial and quality characteristics of the animal food products. However, a potential barrier against the use of insect proteins in animal feed is their public acceptance by consumers. In Western society, the lack of a cultural history of eating insects makes them a novel food. It is noteworthy that available consumer perception surveys showed a high level of support for insects as a protein source in animal feeding, as well as a desire for more information about the topic. However, in order to facilitate consumer acceptance towards the use of insect proteins in animal feeds, it is important that the introduction of this novel source be carried out in a transparent manner. In particular, consumers will have to be consulted and informed throughout the entire production process, in order to avoid the communication bias committed in the past, for example, in the case of protein sources deriving from GMO crops.

References

Agunbiade JA, Adeyemi OA, Ashiru OM et al (2007) Replacement of fish meal with maggot meal in cassava-based layers' diets. J Poultry Sci 44:278–282

Altmann BA, Neumann C, Velten S et al (2018) Meat quality derived from high inclusion of a micro-alga or insect meal as an alternative protein source in poultry diets: a pilot study. Foods 34:1–15

Awoniyi TAM (2007) Health, nutritional and consumers' acceptability assurance of maggot meal inclusion in livestock diet: a review. Int J Trop Med 2:52–56

Ballitoc DA, Sun S (2013) Ground yellow mealworms (Tenebrio molitor L.) feed supplementation improves growth performance and carcass yield characteristics in broilers. Open Science Repository Agriculture, Online (open-access), e23050425. https://doi.org/10.7392/openaccess.23050425

Barrena R, Sánchez M (2012) Neophobia, personal consumer values and novel food acceptance. Food Qual Prefer 27:72–84

Bazoche P, Poret S (2016) What do trout eat: acceptance of insects in animal feed. INRA, pp. 1–14

Belforti M, Gai F, Lussiana C et al (2015) Tenebrio molitor meal in rainbow trout (Oncorhynchus mykiss) diets: effects on animal performance, nutrient digestibility and chemical composition of fillets. Ital J Anim Sci 14:670–675

Biasato I, Gasco L, De Marco M et al (2017) Effects of yellow mealworm larvae (Tenebrio molitor) inclusion in diets for female broiler chickens: implications for animal health and gut histology. Anim Feed Sci Technol 234:253–263

Biasato I, Gasco L, De Marco M et al (2018) Yellow mealworm larvae (Tenebrio molitor) inclusion in diets for male broiler chickens: effects on growth performance, gut morphology and histological findings. Poult Sci 97:540–548

Borgogno M, Dinnella C, Iaconisi V et al (2017) Inclusion of Hermetia illucens larvae meal on rainbow trout (Oncorhynchus mykiss) feed: effect on sensory profile according to static and dynamic evaluations. J Sci Food Agric 97:3402–3411

Bovera F, Loponte R, Marono S et al (2016) Use of Tenebrio molitor larvae meal as protein source in broiler diet: effect on growth performance, nutrient digestibility, and carcass and meat traits. J. Anim. Sci. 94:639–647

Cullere M, Tasoniero G, Giaccone V et al (2016) Black soldier fly as dietary protein source for broiler quails: apparent digestibility, excreta microbial load, feed choice, performance, carcass and meat traits. Animal 10:1923–1930

Cullere M, Tasoniero G, Giaccone V et al (2018) Black soldier fly as dietary protein source for broiler quails: meat proximate composition, fatty acid and amino acid profile, oxidative status and sensory traits. Animal 12:640–647

Cummins VC Jr, Rawles SD, Thompson KR et al (2017) Evaluation of black soldier fly (Hermetia illucens) larvaemeal as partial or total replacement of marine fish meal in practical diets for Pacific white shrimp (Litopenaeus vannamei). Aquaculture 473:337–344

Dobermann D, Swift JA, Field LM (2017) Opportunities and hurdles of edible insects for food and feed. Nutrition Bulletin 42:293–308

Gasco L, Gai F, Maricchiolo G et al (2018) Fish meal alternative protein sources for aquaculture feeds. In: Gasco L, Gai F, Maricchiolo G, Genovese L, Ragonese S, Bottari T, Caruso G (eds) Feeds for the aquaculture sector. Current situation and alternative sources. Springer International Publishing, Berlin, pp 1–28

Henry M, Gasco L, Piccolo G et al (2015) Review on the use of insects in the diet of farmed fish: past and future. Anim Feed Sci Technol 203:1–22

Iaconisi V, Marono S, Parisi G et al (2017) Dietary inclusion of *Tenebrio molitor* larvae meal: effects on growth performance and final quality treats of blackspot sea bream (*Pagellus bogaraveo*). Aquaculture 476:49–58

Iaconisi V, Bonelli A, Pupino R et al (2018) Mealworm as dietary protein source for rainbow trout: body and fillet quality traits. Aquaculture 484:197–204

Józefiak D, Józefiak A, Kierończyk B et al (2016) Insects – a natural nutrient source for poultry – a review. Ann Anim Sci 16:297–313

Khan S, Khan RU, Alam W et al (2018) Evaluating the nutritive profile of three insect meals and their effects to replace soya bean in broiler diet. J Anim Physiol Anim Nutr 102:e662–e668

Laureati M, Proserpio C, Jucker C et al (2016) New sustainable protein sources: consumers' willingness to adopt insects as feed and food. Ital J Food Sci 28:652–668

Leiber F, Gelencsér T, Stamer A et al (2017) Insect and legume-based protein sources to replace soybean cake in an organic broiler diet: effects on growth performance and physical meat quality. Renew Agr Food Syst 32:21–27

Lock ER, Arsiwalla T, Waagbo R (2016) Insect larvae meal as an alternative source of nutrients in the diet of Atlantic salmon (*Salmo salar*) postsmolt. Aquacult. Nutr. 22:1202–1213

Loponte R, Nizza S, Bovera F et al (2017) Growth performance, blood profiles and carcass traits of Barbary partridge (*Allectoris barbara*) fed two different insect larvae meals (*Tenebrio molitor* and *Hermetia illucens*). Res. Vet. Sci. 115:183–188

Makkar HPS, Tran G, Heuze V et al (2014) State-of-the-art on use of insects as animal feed. Anim. Feed Sci. Technol. 197:1–33

Mancini S, Medina I, Iaconisi V et al (2018) Impact of black soldier fly larvae meal on the chemical and nutritional characteristics of rainbow trout fillets. Animal 12:1672–1681

Mancuso T, Baldi L, Gasco L (2016) An empirical study on consumer acceptance of farmed fish fed on insect meals: The Italian case. Aquacult Int 24:1489–1507

Mwaniki Z, Neijat M, Kiarie E (2018) Egg production and quality responses of adding up to 7.5% defatted black soldier fly larvae meal in a corn–soybean meal diet fed to Shaver White Leghorns from wk 19 to 27 of age. Poult. Sci. 0:1–7

Neves ATSG (2015) Determinants of consumers' acceptance of insects as food and feed: a cross – cultural study. Dissertation, Faculty of Sciences, University of Porto.

Onsongo VO, Osuga I, Gachuiri M et al (2018) Insects for income generation through animal feed: effect of dietary replacement of soybean and fish meal with black soldier fly meal on broiler growth and economic performance. J Econ Entomol 11:1966–1973

Panini RL, Pinto SS, Nóbrega RO et al (2017) Effects of dietary replacement of fishmeal by mealworm meal on muscle quality of farmed shrimp *Litopenaeus vannamei*. Food Res Int 102:445–450

Park BS, Um KH, Choi WK et al (2017) Effect of feeding black soldier fly pupa meal in the diet on egg production, egg quality, blood lipid profiles and faecal bacteria in laying hens. Europ Poult Sci 81:1–12

Piccolo G, Iaconisi V, Marono S et al (2017) Effect of *Tenebrio molitor* larvae meal on growth performance, in vivo nutrients digestibility, somatic and marketable indexes of gilthead sea bream (*Sparus aurata*). Anim. Feed Sci. Technol. 226:12–20

Pieterse E, Pretorius Q, Hoffman LC et al (2014) The carcass quality, meat quality and sensory characteristics of broilers raised on diets containing either Musca domestica larvae meal, fish meal or soya bean meal as the main protein source. Animal Production Science 54:622–628

Pieterse E, Erasmus SW, Uushona T et al (2018) Black soldier fly (*Hermetia illucens*) pre-pupae meal as a dietary protein source for broiler production ensures a tasty chicken with standard meat quality for every pot. J. Sci. Food Agric. https://doi.org/10.1002/jsfa.9261

Popoff M, MacLeod M, Leschen W (2017) Attitudes towards the use of insect-derived materials in Scottish salmon feeds. J Ins Food Feed 3:131–138

PROteINSECT (2016) Whitepaper "Insect Protein – feed for the future: addressing the need for feeds of the future today". https://www.allaboutfeed.net/New-Proteins/Articles/2016/4/New-whitepaper-on-insect-protein-for-feed-2796185W/. Accessed 19 Nov 2018.

Renna M, Schiavone A, Gai F et al (2017) Evaluation of the suitability of a partially defatted black soldier fly (*Hermetia illucens* L.) larvae meal as ingredient for rainbow trout (*Oncorhynchus mykiss* Walbaum) diets. J Anim Sci Biotech 8:57

Ruhnke I, Normant C, Campbell DLM et al (2018) Impact of on-range choice feeding with black soldier fly larvae (*Hermetia illucens*) on flock performance, egg quality, and range use of free-range laying hens. Anim Nutr 4:452–460

Sánchez-Muros MJ, Barroso FG, Manzano-Agugliaro F (2014) Insect meal as renewable source of food for animal feeding: a review. J. Clean. Prod. 65:16–27

Sealey WM, Gaylord TG, Barrows FT et al (2011) Sensory analysis of rainbow trout, *Oncorhynchus mykiss*, fed enriched black soldier fly prepupae, *Hermetia illucens*. Journal of the World Aquaculture Society 42:34–45

Secci G, Mancini S, Iaconisi V et al (2018a) Can the inclusion of black soldier fly (*Hermetia illucens*) in diet affect the flesh quality/nutritional traits of rainbow trout (*Oncorhynchus mykiss*) after freezing and cooking? Int J Food Sci Nutr. https://doi.org/10.1080/09637486.2018.1489529

Secci G, Moniello G, Gasco L et al (2018b) Barbary partridge meat quality as affected by *Hermetia illucens* and *Tenebrio molitor* larva meals in feeds. Food Res. Int. 112:291–298

Secci G, Bovera F, Nizza S et al (2018c) Quality of eggs from Lohmann Brown classic laying hens fed black soldier fly meal as substitute for soya bean. Animal. https://doi.org/10.1017/S1751731117003603

Stadtlander T, Stamer A, Buser A et al (2017) *Hermetia illucens* meal as fish meal replacement for rainbow trout on farm. J Insects Food Feed 3:165–175

Turchini G, Moretti VM, Mentasti T et al (2007) Effects of dietary lipid source on fillet chemical composition, flavour volatile compounds and sensory characteristics in the freshwater fish tench (*Tinca tinca* L.). Food Chem. 102:1144–1155

Van Huis A, Oonincx DGAB (2017) The environmental sustainability of insects as food and feed. A review. Agron Sustain Dev 37:43

Veldkamp T, Bosch G (2015) Insects: a protein-rich feed ingredient in pig and poultry diets. Anim Frontiers 5:45–50

Verbeke W, Spranghers T, De Clercq P et al (2015) Insects in animal feed: acceptance and its determinants among farmers, agriculture sector stakeholders and citizens. Anim. Feed Sci. Technol. 204:72–87

Zhou JS, Liu SS, Ji H et al (2018) Effect of replacing dietary fish meal with black soldier fly larvae meal on growth and fatty acid composition of Jian carp (*Cyprinus carpio* var. Jian). Aquacult. Nutr. 24:424–433

Chapter 7
Potential Allergenic Risks of Entomophagy

José Carlos Ribeiro, Luís Miguel Cunha, Bernardo Sousa-Pinto,
and João Fonseca

Abstract Edible insects are a novel food source in the West, prompting the need
for an assessment of their food security risks. One of the major concerns relates to
their allergenic potential, as insects have a close phylogenetic relationship with
crustaceans and house dust mites. Accordingly, several studies have demonstrated
the occurrence of immunologic co-sensitisation between insects and crustaceans/
house dust mites, with tropomyosin and arginine kinase being identified as the
major cross-reacting allergens. This co-sensitisation has been described to be clini-
cally relevant for patients allergic to crustaceans but is still controversial in the case
of individuals allergic to house dust mites. Epidemiological information is still
scarce, with few studies mentioning insects as causative agents of food allergy
(reporting that 0.3–19.4% of food related anaphylactic reactions in Asia were caused
by insects) and case reports lacking in contextual information. Besides food allergy,
insects also present major risks of occupational allergy development through primary sensitisation, although it is not clear which are the responsible allergens.
Therefore, several controversies exist on insects' allergenicity but it is clear that

J. C. Ribeiro · L. M. Cunha (✉)
GreenUPorto & LAQV/REQUIMTE, DGAOT, Faculdade de Ciências da
Universidade do Porto, Campus de Vairão, Vila do Conde, Portugal
e-mail: lmcunha@fc.up.pt

B. Sousa-Pinto
MEDCIDS – Department of Community Medicine, Information and Health
Decision Sciences, Faculty of Medicine, University of Porto, Porto, Portugal

Laboratory of Immunology, Basic and Clinical Immunology Unit, Faculty
of Medicine, University of Porto, Porto, Portugal

CINTESIS – Centre for Health Technology and Services Research, Porto, Portugal

J. Fonseca
MEDCIDS – Department of Community Medicine, Information and Health
Decision Sciences, Faculty of Medicine, University of Porto, Porto, Portugal

CINTESIS – Centre for Health Technology and Services Research, Porto, Portugal

Allergy Unit, CUF Porto Institute & Hospital, Porto, Portugal

© Springer Nature Switzerland AG 2019
G. Sogari et al. (eds.), *Edible Insects in the Food Sector*,
https://doi.org/10.1007/978-3-030-22522-3_7

crustacean-allergic subjects and insect rearing workers are two major risk groups for the development of food and occupational allergy, respectively.

Keywords Edible insects · Novel food · Crustaceans · House dust mites · Food allergy

Introduction

Most studies assessing allergies related to insects have focused on stings from insects belonging to the order *Hymenoptera* (e.g. wasps, bees) (Ludman and Boyle 2015) or inhalant allergy to cockroaches (Pomes et al. 2017). However, given the role of edible insects as novel foods (and, particularly, as novel protein sources), it is also of extreme importance to assess their allergenic potential from the perspective of food security (Belluco et al. 2017). In fact, insects have a close phylogenetic relationship with some common triggers of food and respiratory allergies (Pennisi 2015), namely crustaceans (Loh and Tang 2018) and house dust mites (Calderón et al. 2015). As a result, edible insects can possibly share common allergens (pan-allergens) which may incite cross-reactivity with crustaceans and/or house dust mites (Verhoeckx et al. 2016). On the other hand, allergic reactions can also occur to products derived from insects, such as carmine, a color additive used in the food industry, which is obtained from female *Dactylopius coccus*; however, it is still uncertain if such reactions are triggered by insect-based allergens or due to the low molecular-sized carminic acid acting as an hapten (Müller-Maatsch and Gras 2016).

In this chapter, the molecular mechanisms implied in cross-reactivity/co-sensitisation between edible insects and crustaceans/house dust mites are reviewed. Furthermore, the epidemiology of food allergy to edible insects, and the published case reports and case series describing allergic reactions following the intentional consumption of insects are also described. Lastly, occupational allergy to edible insects and the molecular mechanisms of primary sensitisation are also discussed.

Molecular Mechanisms Underlying Allergy to Edible Insects

Co-Sensitisation/Cross Reactivity Between Edible Insects and Crustaceans/House Dust Mites

Co-sensitisation/cross-reactivity between non-edible insects (e.g. cockroaches) and crustaceans/house dust mites has been thoroughly studied, and arthropod pan-allergens – such as tropomyosin and arginine kinase – have been identified as cross-reacting molecules (Ayuso et al. 2002; Binder et al. 2001). Regarding edible insects with potential application in the food industry, several studies have reported immunologic co-sensitisation between crustaceans/house dust mites and mealworms

(Broekman et al. 2015, 2017a; van Broekhoven et al. 2016; Verhoeckx et al. 2014) crickets (Broekman et al. 2017a; Hall et al. 2018; Srinroch et al. 2015), locusts (Phiriyangkul et al. 2015) and grasshoppers (Broekman et al. 2017a; Leung et al. 1996; Sokol et al. 2017). The main allergens responsible for this co-sensitisation have been identified as tropomyosin (Broekman et al. 2015; Hall et al. 2018; van Broekhoven et al. 2016; Verhoeckx et al. 2014) and arginine kinase (Broekman et al. 2015; Phiriyangkul et al. 2015; Srinroch et al. 2015; Verhoeckx et al. 2014), although other minor arthropod allergens – such as myosin light chain, fructose-biphosphate aldolase, actin, enolase, α-tubulin and β-tubulin – have also been identified (Phiriyangkul et al. 2015; van Broekhoven et al. 2016; Verhoeckx et al. 2014). The role of tropomyosin as a cross-reacting allergen between edible insects and crustaceans/house dust mites has been established with the use of inhibition assays (Leung et al. 1996; Sokol et al. 2017). Tropomyosin from yellow mealworm (*Tenebrio molitor*) has also been shown to have a great amino acid sequence homology with known allergic tropomyosins, including those from arthropods (van Broekhoven et al. 2016).

The role of tropomyosin and arginine kinase as cross-reacting/co-sensitizing allergens has been demonstrated in studies assessing the sera of patients allergic to crustaceans only or to crustaceans and house dust mites. On the other hand, in patients allergic to house dust mites only, neither of these allergens has been identified as cross-reacting/co-sensitizing (van Broekhoven et al. 2016), and there have been conflicting reports about the role of tropomyosin in house dust mite cross reactivity with other arthropods (Bessot et al. 2010; Wong et al. 2016). Instead, other minor arthropod allergens have been identified, such as hexamerin 1B (van Broekhoven et al. 2016), an allergen closely related to hemocyanin (Burmester 2002), which is a shrimp allergen capable of cross-reacting with house dust mites (Faber et al. 2017). Furthermore, hexamerin 1B has also been identified as a co-sensitizing allergen between crustaceans and edible insects such as crickets (Srinroch et al. 2015) or locusts (Phiriyangkul et al. 2015). Other minor arthropod allergens involved in cross-reaction/co-sensitization between edible insects and house dust mites include paramyosin and α-amylase (van Broekhoven et al. 2016).

The influence of processing techniques on the allergenicity of edible insects has also been assessed (Broekman et al. 2015; Hall et al. 2018; Phiriyangkul et al. 2015; van Broekhoven et al. 2016). Overall, co-sensitisation between edible insects and crustaceans does not seem to be significantly altered by thermal processing (Broekman et al. 2015; Phiriyangkul et al. 2015; van Broekhoven et al. 2016), contrarily to what occurs with house dust mites (van Broekhoven et al. 2016), – in fact, tropomyosin of yellow mealworm still maintains its allergenicity after thermal treatments or *in vitro* digestion (van Broekhoven et al. 2016), a behavior that is very similar to crustaceans' tropomyosin (Khan et al. 2018). In addition to resistance to thermal processing or digestion, mealworm tropomyosin presents other important characteristics of food allergens (Bannon 2004), including its abundance on the respective species (Yi et al. 2016), and high sequence homology with known allergy-inducing tropomyosins, which are indications of their allergenic potential.

Concerning allergens from other insects, Hall et al. (2018) assessed the tropomyosin of cricket species *Gryllodes sigillatus,* and found that only a degree of hydrolysis superior to 50% with alcalase was able to eliminate its IgE-binding capacity to shrimp-allergic sera. By contrast, Phiriyangkul et al. (2015) reported that thermal processing was able to significantly alter the intensity and types of allergens identified in co-sensitisation between prawn and Bombay locust – arginine kinase was no longer able to IgE-bind; enolase and hexamerin 1B IgE-binding capacity was diminished; and pyruvate kinase and GADPH (which were not present in the unprocessed extract) appeared as IgE-binding proteins. These results are not surprising considering that it has been reported that arginine kinase is not a thermal stable allergen (Khan et al. 2018).

In vivo studies have supported the clinical significance of cross-reactivity between yellow mealworm and shrimp. Broekman et al. (2016) assessed 15 shrimp or shrimp/house dust mite allergic subjects, and found that the majority (13/15, 86.7%) had a positive double-blind placebo-controlled food challenge with mealworm – this percentage may be considered quite high, especially when considering that the reported risk of reaction between different species of shellfish is about 75% (Sicherer 2001). Furthermore, the authors found that all patients were sensitized to mealworm (having a positive skin prick test), and that the vast majority (14/15) recognized either tropomyosin or arginine kinase, which further highlights the role of these two allergens in cross reactivity/co-sensitisation between edible insects and crustaceans. More studies performing oral challenges (with other species of edible insects besides yellow mealworm) and inhibition assays are necessary to improve the knowledge on cross-reactivity involving edible insects. Even more pressing is the need for research assessing cross-reactivity between house dust mites and edible insects, as studies dealing with patients solely allergic to house dust mites are scarce. In fact, it is still uncertain which are the major allergens that regulate edible insects/house dust mites co-sensitisation, and if house dust mite-allergic patients are at risk of developing clinical food allergy when consuming edible insects.

Primary Sensitisation

Regarding the molecular mechanisms underlying primary sensitisation to allergens of potentially edible insects, most studies have assessed subjects sensitized to silkworm (as a food or inhalant allergen) (Jeong et al. 2016, 2017; Liu et al. 2009; Wang et al. 2016; Zhao et al. 2015; Zuo et al. 2015). While the role of tropomyosin as an allergen is uncertain (Jeong et al. 2017), several other IgE-binding elements have been identified, namely arginine kinase (Liu et al. 2009), chitinase, paramyosin (Zhao et al. 2015), 27-kDa heat-stable glycoprotein (Jeong et al. 2016), thiol peroxiredoxin (Wang et al. 2016), vitellogenin, chitinase, 30 K protein, triosephosphate isomerase, heat shock protein and chymotrypsin inhibitor (Zuo et al. 2015). Arginine kinase, paramyosin and chitinase may play a role in cross-reactivity with other arthropods such as cockroaches, house dust mites or even shrimp due to their high sequence homology with known allergens of these species. In fact, it is reported that there is a high degree of co-sensitisation between silkworm and other common aeroallergens triggers such as house dust mites and cockroaches (Sun et al. 2014).

It is still uncertain if primary sensitisation to insects can lead to cross-reactivity with crustaceans. Linares et al. (2008) described an individual with primary sensitization and respiratory allergy to different species of crickets – the subject had no detectable sIgE to allergic tropomyosins, and had no cross-reactivity for crustaceans or mites. On the other hand, Broekman et al. (2017b) assessed four subjects who were primarily sensitized and had either inhalant or food allergy to yellow mealworm – none of these subjects had a positive oral challenge to shrimp, and only one was sensitized to house dust mites. It was suggested that the major allergens responsible for this primary sensitisation were larval cuticle proteins, which might explain the lack of co-sensitisation with other arthropods. Furthermore, this primary sensitisation and allergy to mealworm can be species-specific because the same subjects showed variability in the degree and percentage of sensitization to different insect species (Broekman et al. 2017a). Of note, species-specific allergy has been reported to other insects, including the housefly (Focke et al. 2003), cockroach (Lehrer et al. 1991), green bottle and bee moth (Siracusa et al. 1994)

Prevalence of Food Allergy to Entomophagy

The prevalence of allergic reactions to insects' consumption has been assessed in two different ways: – (1) self-reported allergic reactions amongst entomophagists (Barennes et al. 2015; Chomchai et al. 2018) and (2) retrospective studies of anaphylactic reactions (Ji et al. 2009; Jiang et al. 2016; Jirapongsananuruk et al. 2007; Piromrat et al. 2008; Rangkakulnuwat et al. 2018).

Studies assessing the prevalence of allergic reactions to the consumption of insects have been performed in Asia, more specifically in Laos provinces outside Vientiane (Barennes et al. 2015), and through an Internet survey in Thailand (Chomchai et al. 2018) (Table 7.1). In the study performed by Barennes et al. (2015) in Laos, the observed prevalence of self-reported allergic reactions was of 7.6% (81/1059), with no reported cases of severe anaphylaxis. On the other hand, in their Internet survey performed in Thailand, Chomchai et al. (2018) observed a frequency of allergic reactions of 12.9% (18/140), of which 22.2% (4/18) reported severe symptoms. In this study, the occurrence of allergic reaction to insects was found to be associated with a history of other allergies, including food allergy to seafood.

Of note, the prevalences reported in these two studies have possibly been overestimated, as allergic reactions were self-reported (Barennes et al. 2015; Chomchai et al. 2018), and as people who suffered allergic reactions may have been more predisposed to participate in the survey (Chomchai et al. 2018).

Despite insects not being mentioned as causative agents of food allergy in several studies assessing the prevalence of food allergy (Lee et al. 2013; Loh and Tang 2018), five different studies performed in China (Ji et al. 2009; Jiang et al. 2016)

Table 7.1 List of studies assessing prevalence of food allergy amongst entomophagists

Reference/ Study	Country	Methodology	Total number of subjects – n	Number of self-reported allergic reactions – n (%)	Species (n)	Other information
Barennes et al. (2015)	Laos	Cross-sectional survey assessing, amongst others, the occurrence of side effects after eating insects	1059	81 (7.6%)	Mostly grasshoppers and stink bugs (number of occurrences not specified)	No subject reported severe anaphylaxis
Chomchai et al. (2018)	Thailand	Internet-based cross-sectional survey of people who practiced entomophagy.	140	18 (12.9%)	Silkworm larva (8–44.4%) Grasshopper (4–22.2%) Cricket (3–16.7%) Bamboo caterpillar (3–16.7%)	Allergic symptoms after insect consumption were associated with a history of respiratory allergy, skin allergy and seafood allergy.

and Thailand (Jirapongsananuruk et al. 2007; Piromrat et al. 2008; Rangkakulnuwat et al. 2018) retrospectively evaluated cases of food anaphylaxis, and have reported on the frequency of events caused by insects (Table 7.2). These studies report a total of 93 cases of food anaphylaxis caused by insects. The species reported to have caused the most reactions were silkworm pupae, crickets and grasshoppers, reflecting the consumption habits in the regions in which these studies were performed. Although not much information is reported about each case, at least one of the subjects that had anaphylaxis following insects' consumption also reported to have suffered from anaphylaxis following shrimp consumption (Piromrat et al. 2008).

Case Reports and Case Series of Food Allergy Caused by Insects

Through literature review, we identified a total of 31 described cases of food allergy caused by consumption of insects. These cases are summarized in Table 7.3, which consists on an adapted update of a table published in our previous article (Ribeiro et al. 2018).

Table 7.2 List of articles which retrospectively analyzed anaphylactic reactions, which included cases caused by insects

Reference	Country	Methodology	Total number of cases of food anaphylaxis – n	Number of cases caused by insects – n (%)	Species (n)	Other information
Ji et al. (2009)	China	Retrospective review of case reports of anaphylactic shock and lethal anaphylaxis caused by food consumption occurring in China between 1980 and 2007.	358	63 (17.6%)	Locust (27) Grasshopper (27) Silkworm pupa (5) Cicada pupa (1) Bee pupa (1) Bee larva (1) Clanis bilineata (1)	Previous history of food allergy was positive in 4 cases, negative in 19 and unknown in 40.
Jiang et al. (2016)	China	Retrospective review of outpatients diagnosed with "anaphylaxis" or "severe allergic reactions" in the Department of Allergy, Peking union medical college hospital, from January 1, 2000 to June 30, 2014	1501	5 (0.3%)	Locusts (2) Cicada (2) Silkworm chrysalis (1)	
Jirapongsananuruk et al. (2007)	Thailand	Retrospective review of medical records of patients admitted with clinical anaphylaxis between January 1, 1999, and December 31, 2004, to Siriraj Hospital, Mahidol University	24	1 (4.2%)	Unknown	

(continued)

Table 7.2 (continued)

Reference	Country	Methodology	Total number of cases of food anaphylaxis – n	Number of cases caused by insects – n (%)	Species (n)	Other information
Piromrat et al. (2008)	Thailand	Retrospective review of all patients who were diagnosed with anaphylaxis at the emergency department (ED) of Bhumibol Adulyadej hospital, Bangkok, from January 1, 2005 to December 31, 2006.	36	7 (19.4%)	Fried grasshoppers and locusts (n not specified)	One woman who presented fried insect-induced anaphylaxis developed shrimp-induced anaphylaxis a few months later
Rangkakulnuwat et al. (2018)	Thailand	Retrospective review of electronic medical records of patients who attended the outpatient and Emergency Departments at Chiang Mai University Hospital from January 2007 to December 2016.	209	17 (8.1%)	Fried insects, namely grasshopper, crickets, silk worms, and bamboo worms (n not specified)	

Table 7.3 List of reported cases of food allergy to consumption of insects

Article	Age/sex/nationality	Species	Clinical symptoms	Personal history	Other characteristics
Freye (1996)	NA/NA/American	Yellow mealworm (*Tenebrio molitor*)	U, D and P (throat, tongue and palm of hands)	Regularly exposed to superworm, yellow mealworm and german cockroach; hypersensitive to inhalation of these species (bronchospasm, conjunctivitis and rhinitis)	Positive SPT and sIgE to superworm, mealworm and cockroach; positive intradermal skin testing to common indoor and outdoor allergens.
Broekman et al. (2017b)	28/M/Dutch	Yellow mealworm (*Tenebrio molitor*)	OA,U	Domestic breeder of yellow mealworm	Positive DBPCFC to mealworm; negative oral challenge to shrimp; sensitized (SPT and ImmunoCAP) to mealworm. IgE binding to tropomyosin, arginine kinase, myosin light and heavy chains and larval cuticle proteins
Broekman et al. (2017b)	32/M/Dutch	Yellow mealworm (*Tenebrio molitor*)	OA,U	Domestic breeder of yellow mealworm; allergic respiratory diseases to mealworm, HDM and birch pollen.	Positive DBPCFC to mealworm; negative oral challenge to shrimp; sensitized (SPT, ImmunoCAP AND ImmunoCAP ISAC) to mealworm and HDM. IgE binding to tropomyosin, arginine kinase, myosin light and heavy chains and larval cuticle proteins
Ji et al. (2008)	37/M/French (visiting China)	Silkworm (*Bombyx mori*) pupa	I (mouth and face), N, S (face), DB, H	History of allergic rhinitis; first time consuming silkworm pupa	

(continued)

Table 7.3 (continued)

Article	Age/sex/nationality	Species	Clinical symptoms	Personal history	Other characteristics
Ji et al. (2008)	21/M/ Chinese	Silkworm (*Bombyx mori*) pupa	P,U,S,H,F,UC	No history of previous allergic shock; first time consuming silkworm pupa	
Ji et al. (2008)	19/M/ Chinese 37/F/Chinese 54/F/Chinese	Silkworm (*Bombyx mori*) pupa	U,S,HA,H,AP,V,DY	No history of previous allergic shock; first time consuming silkworm pupa	
Ji et al. (2008)	M/20/ Chinese	Silkworm (*Bombyx mori*) pupa	P,U,S,H,F,N,UC	No history of previous allergic shock; first time consuming silkworm pupa	
Ji et al. (2008)	5 males: 9–46 3 females: 9–46	Silkworm (*Bombyx mori*) pupa	P,U,S,A,AP,V,N,D	No history of previous allergic shock; no history of previous history; first time consuming silkworm pupa	
Gautreau et al. (2016)	M/44/ Nepalese (living in U.S.A)	Silkworm (*Bombyx mori*) pupa	U,FL,DS	First time consuming silkworm pupa	
Gautreau et al. (2016)	M/18/ Nepalese (living in U.S.A)	Silkworm (*Bombyx mori*) pupa	FL,E (face), U	First time consuming silkworm pupa	
Choi et al. (2010)	F/22/Korean	Vegetable worm (*Cordyceps sinensis*) and silkworm (*Bombyx mori*) pupa	R,C,A	History of allergic rhinitis	
Choi et al. (2010)	M/40/Korean	Vegetable worm (*Cordyceps sinensis*) and silkworm (*Bombyx mori*) pupa	U	History of allergic rhinitis	

Choi et al. (2010)	F/33/Korean	Vegetable worm (*Cordyceps sinensis*) and silkworm (*Bombyx mori*) pupa	U	History of contact urticaria	
Choi et al. (2010)	M/39/Korean	Vegetable worm (*Cordyceps sinensis*) and silkworm (*Bombyx mori*) pupa	P	History of allergic rhinitis and bronchial asthma	
Choi et al. (2010)	F/24/Korean	Vegetable worm (*Cordyceps sinensis*) and silkworm (*Bombyx mori*) pupa	U	History of allergic rhinitis and food allergy to seafood	
Kung et al. (2013)	M/39/Batswana	Mopane worm (*Imbrasia belina*)	I (throat and palatal), U	History of allergic conjunctivitis	Positive SPT to dust mites and mopane worm.
Kung et al. (2011)	M/15/Batswana	Mopane worm (*Imbrasia belina*)	DY, P (palatal), U, S (tongue and lips), HA	History of allergic rhinitis and infantile eczema.	Positive SPT to trees, grass, dust mites, weed, cockroach and mopane worm.
Okezie et al. (2010)	F/36/Batswana	Mopane worm (*Imbrasia belina*)	S (body and face), I (generalized), N, RA, H	Had eaten mopane worm all her life without problem.	
Yew and Ling Kok (2012)	M/46/Chinese	Sago worms (*Rhynchophorus ferrugineus*)	I (face, arms and body), DB, Takotsubo cardiomyopathy	First time eating sago worms	
Piatt (2005)	M/45/American	Cicadas	U	Food allergy to shellfish	

(continued)

Table 7.3 (continued)

Article	Age/sex/ nationality	Species	Clinical symptoms	Personal history	Other characteristics
Sokol et al. (2017)	M/43/ American	Chapulines (*Sphenarium mexicanum*)	I,S (lips and tongue), UC, AP, D	History of allergic rhinoconjunctivitis, bronchial asthma and food allergy to shellfish	Positive SPT and sIgE to grasshopper, chapulines, crickets, cockroach, mites, shellfish, cat and dog. sIgE inhibition with chapulines to grasshopper, crickets, cockroach, mites, shellfish. Identification of tropomyosin in immunoblot.
Sokol et al. (2017)	F/50/ American	Chapulines (*Sphenarium mexicanum*)	I (mouth, throat, generalized), S (face, lips, perioral tissue, throat), DSw, DSp, Sy	History of allergic rhinoconjunctivitis, bronchial asthma, intermittent urticaria, moderately severe atopic dermatitis and food allergy to shellfish	Positive SPT and sIgE to grasshopper, chapulines, crickets, cockroach, mites, shellfish, cat and dog. sIgE inhibition with chapulines to grasshopper, crickets, cockroach and shellfish. Identification of tropomyosin in immunoblot.

F Female, *M* Male, *A* Angioedema, *AP* Abdominal Pain, *C* Conjunctivitis, *D* Diarrhea, *DB* Difficulty in breathing, *DS* Dry Skin, *DSp* Difficulty speaking, *DSw* Difficulty swallowing, *DY* Dyspnea, *E* Edema, *F* Fainting, *FL* Flushed, *H* Hypotension, *HA* Headache, *I* Itchiness, *N* Nausea, *OA* Oral Allergy, *P* Pruritus, *R* Rhinitis, *RA* Rash, *S* Swelling, *Sy* Syncope, *V* Vomiting, *U* Urticaria, *UC* Unconsciousness

Although most of these cases have been reported in Asia and Africa (Choi et al. 2010; Ji et al. 2008; Kung et al. 2011, 2013; Okezie et al. 2010; Yew and Ling Kok 2012), there are also reports of cases occurring in the United States of America (Freye 1996; Gautreau et al. 2016; Piatt 2005; Sokol et al. 2017) and in the Netherlands (Broekman et al. 2017b). The culprit species reflect the consumption habits of the countries in which the reactions occurred - for example, the reported reactions that occurred in China (Ji et al. 2008) were due to silkworm pupae, while the reactions occurring in Botswana were caused by mopane worms (Kung et al. 2011, 2013; Okezie et al. 2010).

Most reported cases were described as occurring after consumption of the edible insect for the first time, which might suggest that the subjects had been sensitized to crustaceans or house dust mites, developing the allergic reaction through cross-reactivity. In fact, in four cases, subjects had a previous history of food allergy to shellfish (Choi et al. 2010; Piatt 2005; Sokol et al. 2017) and, in another 9 cases, subjects were either sensitized to common aeroallergens or had an history of allergic diseases (Broekman et al. 2017b; Choi et al. 2010; Freye 1996; Ji et al. 2008; Kung et al. 2011, 2013).

In three reported cases, reactions appeared to occur through primary sensitization – these cases concerned subjects who had been constantly exposed to yellow mealworm, developing allergy to this species (Broekman et al. 2017b; Freye 1996). Of note, when two of these subjects performed oral challenges to shrimp (Broekman et al. 2017b), neither showed symptoms of allergic reactions.

One of the major limitations when studying food allergy to insects is the lack of reported cases and contextual information about them. This limitation can be due to entomophagy being predominantly common in Asia and Africa, which might result in several reports going unreported or unnoticed. For instance, it is reported that, each year, in China over 1000 patients suffer from anaphylactic reactions following the consumption of silkworm pupa (Ji et al. 2008), but contextual information (e.g. sensitisation profiles and allergic history) about those reactions is mostly unpublished.

Occupational Allergy

Entomologists are an important risk group concerning the development of adverse work-related diseases, such as allergies, venom reactions, infection, infestation and delusional parasitosis (Stanhope et al. 2015). In a similar way to what occurs with environmental exposure to cockroaches (Pomes et al. 2017), the occupational exposure to edible insects also leads to an increased risk of sensitization and development of inhalant and cutaneous allergic diseases (Stanhope et al. 2015).

In fact, there are several reported cases of workers becoming sensitized and developing respiratory and/or skin allergic diseases following the constant exposure to potentially edible insects, including the bee moth (Siracusa et al. 1994, 2003), yellow mealworm (Bernstein et al. 1983; Broekman et al. 2017b; Harris-Roberts et al. 2011; Siracusa et al. 1994, 2003), crickets (Bagenstose Iii et al. 1980; Bartra et al. 2008; Linares et al. 2008), locusts (Burge et al. 1980; Harris-Roberts et al. 2011; Tee et al. 1988), grasshoppers (Lopata et al. 2005; Soparkar et al. 1993) and silkworms (Zuo et al. 2015). In addition, there are cases of subjects allergic to the insects' feed in addition to the insect himself (Bagenstose Iii et al. 1980; Harris-Roberts et al. 2011), and to products derived from the insects such as silk (Uragoda and Wijekoon 1991).

The prevalence of respiratory/cutaneous allergic reactions among workers intensively exposed to insects (Burge et al. 1980; Harris-Roberts et al. 2011; Lopata et al. 2005; Siracusa et al. 2003) varies widely across the different studies performed, ranging from 9% to 60%. This probably reflects studies methodological differences, as well as context specificities. So far, only one study has assessed the possibility of workers sensitized to insects developing food allergy to the species to which they are exposed (Broekman et al. 2017b) – that study described two cases of workers who were sensitized to yellow mealworm and subsequently developed food allergy to that same insect. Therefore, further work is need to comprehend if insect-rearing workers present an increased risk of developing food allergy to insects.

Conclusion

As entomophagy grows in the West, it is necessary to assess possible food risks that might be associated with insects' consumption. One of those risks concerns allergenic potential of insects, not only due to their phylogenetic proximity with crustaceans and house dust mites, but also because edible insects can function as sources of allergens themselves. Crustacean-allergic subjects appear to be at risk of food allergy to insects, through cross-reactivity mainly mediated by tropomyosin or arginine kinase. It is still uncertain if house dust mites-allergic subjects can also develop food allergic reaction to insects by cross-reactive mechanisms, and what are the allergens that regulate co-sensitisation between them. Insect-rearing workers are also at risk of developing occupational respiratory and/or skin allergic diseases through primary sensitisation to insects. This sensitisation can also lead to food allergy to the same species workers had been sensitized.

References

Ayuso R, Reese G, Leong-Kee S, Plante M, Lehrer SB (2002) Molecular basis of arthropod cross-reactivity: Ige-binding cross-reactive epitopes of shrimp, house dust mite and cockroach tropomyosins. Int Arch Allergy Immunol 129(1):38–48. https://doi.org/10.1159/000065172
Bagenstose Iii AH, Mathews KP, Homburger HA, Saaveard-Delgado AP (1980) Inhalant allergy due to crickets. J Allergy Clin Immunol 65(1):71–74. https://doi.org/10.1016/0091-6749(80)90180-3
Bannon GA (2004) What makes a food protein an allergen? Curr Allergy Asthma Rep 4(1):43–46

Barennes H, Phimmasane M, Rajaonarivo C (2015) Insect consumption to address undernutrition, a national survey on the prevalence of insect consumption among adults and vendors in Laos. PLoS One 10(8). https://doi.org/10.1371/journal.pone.0136458

Bartra J, Carnes J, Munoz-Cano R, Bissinger I, Picado C, Valero AL (2008) Occupational asthma and rhinoconjunctivitis caused by cricket allergy. J Investig Allergol Clin Immunol 18(2):141–142

Belluco S, Halloran A, Ricci A (2017) New protein sources and food legislation: the case of edible insects and EU law. Food Security 9(4):803–814. https://doi.org/10.1007/s12571-017-0704-0

Bernstein DI, Gallagher JS, Leonard Bernstein I (1983) Mealworm asthma: clinical and immunologic studies. J Allergy Clin Immunol 72(5 PART 1):475–480. https://doi.org/10.1016/0091-6749(83)90584-5

Bessot JC, Metz-Favre C, Rame JM, De Blay F, Pauli G (2010) Tropomyosin or not tropomyosin, what is the relevant allergen in house dust mite and snail cross allergies? Eur Ann Allergy Clin Immunol 42(1):3–10

Binder M, Mahler V, Hayek B, Sperr WR, Schöller M, Prozell S et al (2001) Molecular and immunological characterization of arginine kinase from the indianmeal moth, Plodia interpunctella, a novel cross-reactive invertebrate pan-allergen. J Immunol 167(9):5470–5477

Broekman H, Knulst A, den Hartog Jager S, Monteleone F, Gaspari M, de Jong G et al (2015) Effect of thermal processing on mealworm allergenicity. Mol Nutr Food Res 59(9):1855–1864. https://doi.org/10.1002/mnfr.201500138

Broekman H, Verhoeckx KC, Jager CFD, Kruizinga AG, Pronk-Kleinjan M, Remington BC et al (2016) Majority of shrimp-allergic patients are allergic to mealworm. J Allergy Clin Immunol 137(4):1261. https://doi.org/10.1016/j.jaci.2016.01.005

Broekman H, Knulst AC, de Jong G, Gaspari M, Jager CFD, Houben GF, Verhoeckx KCM (2017a) Is mealworm or shrimp allergy indicative for food allergy to insects? Mol Nutr Food Res 61(9). https://doi.org/10.1002/mnfr.201601061

Broekman H, Knulst AC, Jager CFD, van Bilsen JHM, Raymakers FML, Kruizinga AG et al (2017b) Primary respiratory and food allergy to mealworm. J Allergy Clin Immunol 140(2):600. https://doi.org/10.1016/j.jaci.2017.01.035

Burge PS, Edge G, O'Brien IM, Harries MG, Hawkins R, Pepys J (1980) Occupational asthma in a research centre breeding locusts. Clin Allergy 10(4):355–363

Burmester T (2002) Origin and evolution of arthropod hemocyanins and related proteins. J Comp Physiol B 172(2):95–107

Calderón MA, Linneberg A, Kleine-Tebbe J, De Blay F, Hernandez Fernandez de Rojas D, Virchow JC, Demoly P (2015) Respiratory allergy caused by house dust mites: what do we really know? J Allergy Clin Immunol 136(1):38–48. https://doi.org/10.1016/j.jaci.2014.10.012

Choi GS, Shin YS, Kim JE, Ye YM, Park HS (2010) Five cases of food allergy to vegetable worm (Cordyceps sinensis) showing cross-reactivity with silkworm pupae. Allergy: European Journal of Allergy and Clinical Immunology 65(9):1196–1197. https://doi.org/10.1111/j.1398-9995.2009.02300.x

Chomchai S, Laoraksa P, Virojvatanakul P, Boonratana P, Chomchai C (2018) Prevalence and cluster effect of self-reported allergic reactions among insect consumers: an internet survey from Thailand. Asian Pac J Allergy Immunol. https://doi.org/10.12932/ap-220218-0271

Faber MA, Pascal M, El Kharbouchi O, Sabato V, Hagendorens MM, Decuyper II et al (2017) Shellfish allergens: Tropomyosin and beyond. Allergy 72(6):842–848. https://doi.org/10.1111/all.13115

Focke M, Hemmer W, Wohrl S, Gotz M, Jarisch R, Kofler H (2003) Specific sensitization to the common housefly (Musca domestica) not related to insect panallergy. Allergy 58(5):448–451

Freye HB (1996) Anaphylaxis to the ingestion and inhalation of Tenebrio molitor (mealworm) and Zophobas morio (superworm). Allergy Asthma Proc 17(4):215–219

Gautreau M, Restuccia M, Senser K, Weisberg SN (2016) Familial anaphylaxis after silkworm ingestion. Prehosp Emerg Care:1–3. https://doi.org/10.1080/10903127.2016.1204035

Hall F, Johnson PE, Liceaga A (2018) Effect of enzymatic hydrolysis on bioactive properties and allergenicity of cricket (Gryllodes sigillatus) protein. Food Chem 262:39–47. https://doi.org/10.1016/j.foodchem.2018.04.058

Harris-Roberts J, Fishwick D, Tate P, Rawbone R, Stagg S, Barber CM, Adisesh A (2011) Respiratory symptoms in insect breeders. Occupational Medicine (London) 61(5):370–373. https://doi.org/10.1093/occmed/kqr083

Jeong KY, Son M, Lee JY, Park KH, Lee JH, Park JW (2016) Allergenic characterization of 27-kDa glycoprotein, a novel heat stable allergen, from the Pupa of silkworm, Bombyx mori. J Korean Med Sci 31(1):18–24. https://doi.org/10.3346/jkms.2016.31.1.18

Jeong KY, Han IS, Lee JY, Park KH, Lee JH, Park JW (2017) Role of tropomyosin in silkworm allergy. Mol Med Rep 15(5):3264–3270. https://doi.org/10.3892/mmr.2017.6373

Ji KM, Zhan ZK, Chen JJ, Liu ZG (2008) Anaphylactic shock caused by silkworm pupa consumption in China. Allergy (European Journal of Allergy and Clinical Immunology) 63(10):1407–1408. https://doi.org/10.1111/j.1398-9995.2008.01838.x

Ji K, Chen J, Li M, Liu Z, Wang C, Zhan Z et al (2009) Anaphylactic shock and lethal anaphylaxis caused by food consumption in China. Trends Food Sci Technol 20(5):227–231. https://doi.org/10.1016/j.tifs.2009.02.004

Jiang N, Yin J, Wen L, Li H (2016) Characteristics of anaphylaxis in 907 chinese patients referred to a tertiary allergy center: a retrospective study of 1,952 episodes. Allergy Asthma Immunol Res 8(4):353–361. https://doi.org/10.4168/aair.2016.8.4.353

Jirapongsananuruk O, Bunsawansong W, Piyaphanee N, Visitsunthorn N, Thongngarm T, Vichyanond P (2007) Features of patients with anaphylaxis admitted to a university hospital. Ann Allergy Asthma Immunol 98(2):157–162. https://doi.org/10.1016/s1081-1206(10)60689-8

Khan MU, Ahmed I, Lin H, Li Z, Costa J, Mafra I et al (2018) Potential efficacy of processing technologies for mitigating crustacean allergenicity. Crit Rev Food Sci Nutr:1–24. https://doi.org/10.1080/10408398.2018.1471658

Kung SJ, Fenemore B, Potter PC (2011) Anaphylaxis to Mopane worms (Imbrasia belina). Ann Allergy Asthma Immunol 106(6):538–540. https://doi.org/10.1016/j.anai.2011.02.003

Kung SJ, Mazhani L, Steenhoff AP (2013) Allergy in Botswana. Current Allergy and Clinical Immunology 26(4):202–209

Lee AJ, Thalayasingam M, Lee BW (2013) Food allergy in Asia: how does it compare? Asia Pac Allergy 3(1):3–14. https://doi.org/10.5415/apallergy.2013.3.1.3

Lehrer SB, Horner WE, Menon PK, Oliver J, Hauck P (1991) Cockroach allergenic activity: Analysis of commercial cockroach and dust extracts. J Allergy Clin Immunol 88(6):895–901

Leung PSC, Wing Kuen C, Duffey S, Hoi Shan K, Gershwin ME, Ka Hou C (1996) IgE reactivity against a cross-reactivity allergen in crustacea and mollusca: evidence for tropomyosin as the common allergen. J Allergy Clin Immunol 98(5):954–961

Linares T, Hernandez D, Bartolome B (2008) Occupational rhinitis and asthma due to crickets. Ann Allergy Asthma Immunol 100(6):566–569

Liu Z, Xia L, Wu Y, Xia Q, Chen J, Roux KH (2009) Identification and characterization of an arginine kinase as a major allergen from silkworm (Bombyx mori) larvae. Int Arch Allergy Immunol 150(1):8–14. https://doi.org/10.1159/000210375

Loh W, Tang MLK (2018) The epidemiology of food allergy in the global context. Int J Environ Res Public Health 15(9). https://doi.org/10.3390/ijerph15092043

Lopata AL, Fenemore B, Jeebhay MF, Gade G, Potter PC (2005) Occupational allergy in laboratory workers caused by the African migratory grasshopper Locusta migratoria. Allergy 60(2):200–205. https://doi.org/10.1111/j.1398-9995.2005.00661.x

Ludman SW, Boyle RJ (2015) Stinging insect allergy: current perspectives on venom immunotherapy. J Asthma Allergy 8:75–86. https://doi.org/10.2147/JAA.S62288

Müller-Maatsch, J., & Gras, C. (2016). The "Carmine Problem" and Potential Alternatives. In R. Carle & R. Schweiggert (Eds.), Handbook on Natural Pigments in Food and Beverages: Industrial Applications for Improving Food Color (pp. 385–428). Cambridge, United Kingdom: Elsevier

Okezie OA, Kgomotso KK, Letswiti MM (2010) Mopane worm allergy in a 36-year-old woman: a case report. J Med Case Reports 4. https://doi.org/10.1186/1752-1947-4-42

Pennisi E (2015) All in the (bigger) family: revised arthropod tree marries crustacean and insect fields. Sci Total Environ 347:220–221

Phiriyangkul P, Srinroch C, Srisomsap C, Chokchaichamnankit D, Punyarit P (2015) Effect of food thermal processing on allergenicity proteins in Bombay locust (Patanga Succincta). Int J Food Eng 1(1):23–28. https://doi.org/10.18178/ijfe.1.1.23-28

Piatt JD (2005) Case report: urticaria following intentional ingestion of cicadas. Am Fam Physician 71(11):2048. 2050

Piromrat K, Chinratanapisit S, Trathong S (2008) Anaphylaxis in an emergency department: A 2-year study in a tertiary-care hospital. Asian Pac J Allergy Immunol 26(2–3):121–128

Pomes A, Mueller GA, Randall TA, Chapman MD, Arruda LK (2017) New insights into cockroach allergens. Curr Allergy Asthma Rep 17(4):25. https://doi.org/10.1007/s11882-017-0694-1

Rangkakulnuwat P, Sutham K, Lao-Araya M (2018) Anaphylaxis: Ten-year retrospective study from a tertiary-care hospital in Northern Thailand. Asian Pac J Allergy Immunol. https://doi.org/10.12932/ap-210318-0284

Ribeiro JC, Cunha LM, Sousa-Pinto B, Fonseca J (2018) Allergic risks of consuming edible insects: a systematic review. Mol Nutr Food Res 62(1). https://doi.org/10.1002/mnfr.201700030

Sicherer SH (2001) Clinical implications of cross-reactive food allergens. J Allergy Clin Immunol 108(6):881–890. https://doi.org/10.1067/mai.2001.118515

Siracusa A, Bettini P, Bacoccoli R, Severini C, Verga A, Abbritti G (1994) Asthma caused by live fish bait. J Allergy Clin Immunol 93(2):424–430. https://doi.org/10.1016/0091-6749(94)90350-6

Siracusa A, Marcucci F, Spinozzi F, Marabini A, Pettinari L, Pace ML, Tacconi C (2003) Prevalence of occupational allergy due to live fish bait. Clin Exp Allergy 33(4):507–510. https://doi.org/10.1046/j.1365-2222.2003.01641.x

Sokol WN, Wünschmann S, Agah S (2017) Grasshopper anaphylaxis in patients allergic to dust mite, cockroach, and crustaceans: is tropomyosin the cause? Ann Allergy Asthma Immunol 119(1):91–92. https://doi.org/10.1016/j.anai.2017.05.007

Soparkar GR, Patel PC, Cockcroft DW (1993) Inhalant atopic sensitivity to grasshoppers in research laboratories. J Allergy Clin Immunol 92(1 Pt 1):49–55. https://doi.org/10.1016/0091-6749(93)90036-F

Srinroch C, Srisomsap C, Chokchaichamnankit D, Punyarit P, Phiriyangkul P (2015) Identification of novel allergen in edible insect, Gryllus bimaculatus and its cross-reactivity with Macrobrachium spp. allergens. Food Chem 184:160–166. https://doi.org/10.1016/j.foodchem.2015.03.094

Stanhope J, Carver S, Weinstein P (2015) The risky business of being an entomologist: a systematic review. Environ Res 140:619–633. https://doi.org/10.1016/j.envres.2015.05.025

Sun, B. Q., Zheng, P. Y., Wei, N. L., Huang, H. M., & Zeng, G. Q. (2014). Co-Sensitization to Silkworm Moth (Bombyx mori) and 9 Inhalant Allergens among Allergic Patients in Guangzhou, Southern China. PLoS ONE, 9. doi:10.1371/journal.pone.0094776

Tee RD, Gordon DJ, Hawkins ER, Nunn AJ, Lacey J, Venables KM et al (1988) Occupational allergy to locusts: an investigation of the sources of the allergen. J Allergy Clin Immunol 81(3):517–525

Uragoda CG, Wijekoon PN (1991) Asthma in silk workers. J Soc Occup Med 41(3):140–142

van Broekhoven S, Bastiaan-Net S, de Jong NW, Wichers HJ (2016) Influence of processing and in vitro digestion on the allergic cross-reactivity of three mealworm species. Food Chem 196:1075–1083. https://doi.org/10.1016/j.foodchem.2015.10.033

Verhoeckx K, van Broekhoven S, den Hartog-Jager CF, Gaspari M, de Jong GAH, Wichers HJ et al (2014) House dust mite (Der p 10) and crustacean allergic patients may react to food containing Yellow mealworm proteins. Food Chem Toxicol 65:364–373. https://doi.org/10.1016/j.fct.2013.12.049

Verhoeckx K, Broekman H, Knulst A, Houben G (2016) Allergenicity assessment strategy for novel food proteins and protein sources. Regul Toxicol Pharmacol 79:118–124. https://doi.org/10.1016/j.yrtph.2016.03.016

Wang H, Hu W, Liang Z, Zeng L, Li J, Yan H et al (2016) Thiol peroxiredoxin, a novel allergen from Bombyx mori, modulates functions of macrophages and dendritic cells. Am J Transl Res 8(12):5320–5329

Wong L, Huang CH, Lee BW (2016) Shellfish and house dust mite allergies: is the link tropomyosin? Allergy Asthma Immunol Res 8(2):101–106. https://doi.org/10.4168/aair.2016.8.2.101

Yew KL, Ling Kok VS (2012) Exotic food anaphylaxis and the broken heart: sago worm and takotsubo cardiomyopathy. Med J Malaysia 67(5):540–541

Yi L, Van Boekel MAJS, Boeren S, Lakemond CMM (2016) Protein identification and in vitro digestion of fractions from Tenebrio molitor. Eur Food Res Technol 242(8):1285–1297. https://doi.org/10.1007/s00217-015-2632-6

Zhao X, Li L, Kuang Z, Luo G, Li B (2015) Proteomic and immunological identification of two new allergens from silkworm (Bombyx mori L.) pupae. Central European Journal of Immunology 40(1):30–34. https://doi.org/10.5114/ceji.2015.50830

Zuo J, Lei M, Yang R, Liu Z (2015) Bom m 9 from Bombyx mori is a novel protein related to asthma. Microbiol Immunol 59(7):410–418. https://doi.org/10.1111/1348-0421.12271

Chapter 8
Insects as Food: The Legal Framework

Francesca Lotta

Abstract Even though entomophagy is a very old practice, it is considered a new culinary phenomenon in most Western Countries and, as such, it has received little attention from legislators.

In the European Union, the regulatory status of insects has been quite controversial until the adoption of the new novel food regulation. In the old novel food regulation no mention was expressly made to insects as novel food and this resulted in different approaches of the European Member States. In some Member States whole insects and their parts were considered outside the scope of the novel food regulation and their placing on the market was not subject to pre-market authorization while other Member States considered insects as a novel food and as such subject to the risk assessment procedure provided by the law.

Through the adoption of the new novel food regulation, the legal status of edible insects has been clarified: insects and their part now fall in the definition of novel food and they need to be authorized before being placed on the market. Beside the authorization process, the classification of insects as food poses new challenges when it comes to the legislation applicable to insects farming, slaughtering and processing.

In the United States, the approach was not different: the Food and Drug Administration (FDA) has devoted significant attention to insects in human food as defects, but has given little public attention to insects as human food or as an intentional component of human food. Their regulatory classification is therefore still unclear since they shall either be approved as food additive or their use shall be generally recognized as safe (GRAS) to be legally placed on the US market.

Keywords Novel food · Authorization procedures · European food safety authority · Regulatory classification · GRAS status

F. Lotta (✉)
Ikea Food Services AB, Malmö, Sweden
e-mail: francesca.lotta@ikea.com

© Springer Nature Switzerland AG 2019
G. Sogari et al. (eds.), *Edible Insects in the Food Sector*,
https://doi.org/10.1007/978-3-030-22522-3_8

Introduction

The term entomophagy refers to the practice of eating insects by humans. Insects represent a traditional food category in many cultures of the world and they have played an important part in the history of human nutrition in Africa, Asia and Latin America (Bodenheimer 1951). It has been estimated that entomophagy is practiced in at least 113 countries with over 2000 documented edible insect species (Jongema 2017).

Despite their wide consumption around the world and the benefits related to their consumption (Dobermann et al. 2017), insects are a new culinary phenomenon in Western Countries and their regulatory classification as food has been quite controversial.

This paper aims to provide an overview of the legislative framework applicable to edible insects in two of the biggest Western markets: the European Union and the United States of America. The document will analyze their regulatory classification and the rules food operators need to comply with to legally place them on the market.

European Union: The Regulatory Status of Insects as Food Before Regulation (EU) 2283/2015

At European level, edible insects have been subject to specific legislation only through the adoption of the new *novel food* regulation.[1] The reason for this shift is due to the growing interest in edible insects as sustainable source of proteins as well as harmonization in relation to the sale of edible insects across all European member States.

Before the adoption of the new *novel food* regulation, there was legal uncertainly on the regulatory classification of edible insects. It should be noted that, except for Regulation (EC) 834/2007 on organic production, in the European legislation no mention was expressly made to insects as food.[2] Notwithstanding this legal gap, insects and insect-based products were usually considered to fall within the scope of Regulation (EC) 258/1997 on novel foods[3] (here in after "old *novel food* Regulation"), although this qualification was far to be undisputed (Paganizza 2016).

---

[1] Regulation (EU) 2015/2283 of the European Parliament and of the Council of 25 November 2015 on novel foods, amending Regulation (EU) No 1169/2011 of the European Parliament and of the Council and repealing Regulation (EC) No 258/97 of the European Parliament and of the Council and Commission Regulation (EC) No 1852/2001, L 327/1.

[2] According to the Food and Agriculture Organization (FAO), *"the absence of clear legislation and norms guiding the use of insects as food and feed is among the major limiting factors hindering the industrial development of farming insects to supply the food and feed sectors"* (FAO 2013).

[3] Regulation (EC) of the European Parliament and of the Council of 27 January 1997 concerning novel foods and novel food ingredients, OJ L 43, 14.2.1997.

According to the definition set forth in the old *novel food* Regulation, to be classified as novel, foods or food ingredients need to comply with a twofold condition: (a) they shall not been used for human consumption to a significant degree within the Community before 15 May 1997; and (b) they shall fall within one of categories set forth by Article 1(2) of the old *novel food* regulation.[4] While it was a matter of fact that within the European Union, people have not been consuming insects over the past 50 years, the fulfillment of the second condition was disputed since insects did not seem to fall within any category provided under Article 1(2). They seemed in some extend to fall within the category (e) which gathered "*foods and food ingredients consisting of or isolated from plants and food ingredients isolated from animals*" even though a strict interpretation of the norm would lead to consider as novel only food ingredients isolated from insects, and not insects as such.

The main consequence of the *novel food* status is that *novel foods* are subject to a pre-market authorization aimed at assessing their safety in consideration to the expected use before being placed on the market.

The Different Approach of the European Member States

Lacking clarity in the legislation, the European Member States have interpreted the old novel food regulation in different ways, which has resulted in varied approaches to this topic. In some Member States whole insects have been considered outside the scope of the old *novel food* regulation and their placing on the market has not been subject to pre-market authorization, while other Member States have considered both whole insects and their parts as a novel food and as such subject to the risk assessment procedure provide by the law.

Belgium, United Kingdom and the Netherlands are the leading example of the first group of Countries.

In 2014, the Belgian Federal Agency for the Safety of the Food Chain (FASFC) and the Health High Council issued an opinion in which they have assessed the microbiological, chemical and physical hazards related to the use of insects (FASFC 2014). The document was followed by a circular, which has identified the species of insects which commercialization is possible in Belgium without the submission of a novel food application as well as the conditions for their commercialization (FASFC 2016). The circular clearly states that this regime does not apply to the

---

[4]Article 1(2) provides the following categories (a) foods and food ingredients containing or consisting of genetically modified organisms within the meaning of Directive 90/220/EEC; (b) foods and food ingredients produced from, but not containing, genetically modified organisms; (c) foods and food ingredients with a new or intentionally modified primary molecular structure; (d) foods and food ingredients consisting of or isolated from micro-organisms, fungi or algae; (e) foods and food ingredients consisting of or isolated from plants and food ingredients isolated from animals, except for foods and food ingredients obtained by traditional propagating or breeding practices and having a history of safe food use.

ingredients isolated from insects (e.g. insect proteins) and to insects imported from third countries: in these cases a novel food application is required to place the product on the market.

In the United Kingdom the selling of insects has been tolerated by the authorities to the extent that in July 2015, the Government invited food business operators to submit data aimed at demonstrating a history of safe consumption before 1997. In particular, insect business operators have been required to submit any kind of evidences (e.g. comprehensive sales information, personal testimonies and import/export information) to exclude the *novel food* status of edible insects and insect-based products. Unfortunately, the amount of information supplied in response to the inquiry was not considered solid enough to demonstrate a history of safe consumption in the United Kingdom.

The Netherlands is usually regarded as the Country in Europe with the highest level of tolerance toward the practice of entomophagy (Paganizza 2016, 32). In 2014, upon request of the Netherlands Food and Consumer Product Safety Authority, the Office for Risk Assessment & Research (ORAR) published a document on the chemical, microbiological and parasitological risks related to the consumption of insects (ORAR 2014). The analysis was limited to the insect species reared for human consumption in the Netherlands such as the mealworm beetle (*Tenebrio molitor*), the lesser mealworm beetle (*Alphitobius diaperinus*) and the European migratory locust (*Locusta migratoria*).

According to the findings of the Office for Risk Assessment, insects shall be subject to the same hygiene criteria applicable to meat preparations and the chemical, microbiological and parasitological risk related to their consumption can be controlled through the use of adequate production methods. Insects can cause allergic reactions to sensitive individuals and, due the presence of chitin, the expected daily intake of dried or freeze dried insects shall not exceeds 45 grams per day (ORAR 2014, 3).

A completely different approach to the topic was taken in other European Countries such as Italy and Sweden.

In 2013, the Italian Ministry of Health (Ministero della Salute) published an explanatory note in which it clearly stated that since edible insects are products of animal origin lacking a history of safe consumption in the European Union before 1997, they fall into the scope of the *novel food* regulation. It follows that their placing on the market is subject to pre-marketing approval unless the food business operator is able to demonstrate a long history of consumption either submitting the documentation or providing an official statement of an European Member State Authority that certifies that the insect species has a history of safe consumption in its Country (Ministero della Salute 2013).

In 2017, the Swedish National Food Agency (Livsmedelsverke) issued a press release stating that edible insects fall into the scope of the novel food regulation. The aim of the legislation is to protect consumers from unknown risks such as allergic reaction and food poisoning. Food companies that wish to place on the Swedish market insect-based products shall demonstrate the absence of these risks through the submission of a scientific dossier (Livsmedelsverke 2017).

Authorizing Insects as Novel Food Under Regulation 2283/2015/EU

The growing interest in using insects as alternatives to mainstream animal sources and the reported benefits related to their consumption as food (Testa et al. 2016) has led the European Commission to require the European Food Safety Authority (EFSA) to assess the microbiological, chemical and environmental risks arising from the production and consumption of insects as food and feed. According to the EFSA's scientific opinion, the microbiological, chemical and environmental hazards related to the consumption of insects vary according to the insect species, substrate used, stage of harvesting and processing (EFSA 2015). Even though more research is needed due to the lack of detailed information about the magnitude and frequency of the use of insects as food and feed in Europe, the overall opinion of the Authority is that the risks of using insects as food or feed are no greater than those associated with other animals (Finke et al. 2015, 247).

Several months after the issuing of the EFSA's opinion, the European Commission published the new novel food regulation which clarifies the legal status of insects and sets forth harmonized rules for their placing on the market in the European Union. According to Recital (8) of the new *novel food* Regulation, the notion of novel food needs to be revised to take into consideration the scientific and technological developments that have occurred since 1997 and shall also include whole insects and their parts.

Under the new novel food Regulation, insects and their parts fall within category (v) of Article 3(2) which encompasses food consisting of, isolated from or produced from animals or their parts. The regulatory classification of insects as novel food implies the need for authorization before being placed on the market unless the food business operator is able to demonstrate that the insect species was used for human consumption to a significant degree within the Union before 15 May 1997.

The procedure to assess the novel food status of a substance is currently set forth in the Commission implementing Regulation (EU) 2018/456.[5] Previously, the assessment of the novel food status was performed on an informal and anonymous basis by the national authorities, Regulation (EU) 2018/456 establishes a procedure that requires the food business operator to disclose sensitive information such as the production method and the flowchart. Even though under the implementing regulation, the food business operator might require that some information shall be treated as confidential, the amount of information required to perform the assessment may discourage food business operators to use it and opt directly for a novel food application.

---

[5] Commission implementing Regulation (EU) 2018/456 on the procedural steps of the consultation process for determination of novel food status in accordance with Regulation (EU) 2015/2283 of the European Parliament and the Council on novel foods, L 77/6 of 20.2.2018.

The new novel food regulation provides two procedures to authorize the placing on the market of a *novel food* substance: a general procedure and a notification procedure for traditional foods from third Countries (Pisanello et al. 2018).

The general procedure, set forth in Articles 10–13 of the Regulation, applies to any type of novel food and it takes at least 17 months. The food business operator is required to submit a scientific dossier which is evaluated by the Commission with the support – if deemed necessary – of the European Food Safety Authority. In the novel food application, drafted according to Regulation (EU) 2017/2469[6] and the guidelines provided by EFSA (2016a, 2018) the applicant shall take in consideration the insect species, the substrate used as well as the methods for farming and processing (EFSA 2016a, b).

The notification procedure is provided under Article 14-20 and can be used for placing on the market a traditional food from a third country which fulfill a twofold condition: (i) it has been consumed in a third country for at least 25 years as a part of the customary diet of a significant number of people and (ii) it is derived from primary production as defined in Regulation (EC) No 178/2002, regardless of whether or not it is processed or unprocessed foods. This procedure is faster than the general one since the applicant is not required to prove the safety of the product but only the history of safe use in a third Country. On the other side, the applicant that decides to use the notification procedure cannot benefit of the data protection provided under Article 26 of the new novel food Regulation when specific conditions are met.

Following the positive completion of the procedure, the novel food product is included – through an implementing act issued by the European Commission – in the novel food list established by Regulation (EU) 2017/2470.[7] The authorizations are very specific and any change affecting product specifications, production methods (such as the type of substrate used for feeding insects) and conditions of use would require a new application.

The Placing on the Market of Insects During the Transitional Period

Following the entering into force of the new novel food Regulation, insects and insect-based products also need to be authorized in the Countries where their commercialization was possible under the old regime (i.e. UK, Netherlands and

---

[6] Commission Implementing Regulation (EU) 2017/2469 of 20 December 2017 laying down administrative and scientific requirements for applications referred to in Article 10 of Regulation (EU) 2015/2283 of the European Parliament and of the Council on novel foods, L 351/64 of 30.12. 2017.

[7] Commission Implementing Regulation (EU) 2017/2470 of 20 December 2017 establishing the Union list of novel food in accordance with Regulation (EU) 2015/2283 of the European Parliament and of the Council on novel foods, L 351/72 of 30.12.2017.

Belgium). Since the completion of a *novel food* application might require up to 17 months, the European legislator has set forth specific provisions to enable food business operators to continue placing on the market insects and insect products until the authorization is granted.

Article 35(2) of the new *novel food* regulation provides that foods not falling within the scope of the old novel food regulation, which are lawfully placed on the market by January 1st, 2018 and which fall within the scope of the new *novel food* regulation may continue to be placed on the market until a decision is taken following an application for authorisation of a novel food or a notification of a traditional food from a third country submitted by the date specified in the implementing rules adopted in accordance with Article 13 or 20 of this Regulation respectively, but no later than 2 January 2020.

In the light of this provision, food business operators which have lawfully placed on the market insects and insect products before January 1st, 2018 may continue to sell their products under the new *novel food* regulation, provided they submit an application or notification for the products before January 1, 2019 as established in Regulation (EU) 2017/2469. After January 1, 2019 only such insect species and food produced from those species for which an application for novel food has been submitted to the European Commission by that date can be marketed.

So far, according to the EU Commission summary of the ongoing applications, there are six pending applications concerning edible insects: house cricket (*Acheta domesticus*), whole and ground lesser mealworm (*Alphitobius diaperinus*) larvae products; dried crickets (*Gryllodes sigillatus*); migratory locust (*Locusta migratoria*), dried mealworm (*Tenebrio molitor*), mealworm (*Tenebrio molitor*).[8]

Insects from Farm to Fork: The Regulatory Framework

According to the opinion provided by the European Food Safety Authority (EFSA), both biological and chemical hazards related to the consumption of insects are strongly affected by the type of insect species, the stage of harvesting, the specific production methods and the type of substrate used to feed them (EFSA 2015, 1).

Even in the Countries in which the marketing of insects is not restricted due to a more flexible interpretation of the old novel food regulation, the production and the sale of insects is limited to the insect species which have been subject to a specific assessment by the national authorities.[9] With the entering into force of the new novel food regulation, only the insect species which food safety has been assessed following one of the procedures set forth in the Regulation could be placed into the European market.[10]

---

[8] The list of the pending applications is available at https://ec.europa.eu/food/safety/novel_food/authorisations/summary-ongoing-applications-and-notifications_en, last accessed on 21.12.2018.

[9] See par. 1.2.

[10] Except for the insect-products under the scope of transition rules.

Insect rearing and harvesting is not specifically regulated at European level (Lähteenmäki-Uutela et al. 2017) but some European Authorities have published guidelines aimed at supporting both authorities and insect business operators (EVIRA 2018; FASFC 2018).

Edible insects shall be qualified as "farmed animals" and, as such, subject to the same rules applicable to the other livestock.[11] Insects can be fed only with safe feed listed in the Catalogue of feed material[12]: the use of ruminant proteins, catering waste, meat-and-bone meal and manure is prohibited as well as the use of feces for animal nutritional purposes.[13]

The restrictions applicable to feed materials pose interesting questions on the feasibility to use food waste as insects 'feed material'. Studies has shown that, due to its availability in large quantities and its low cost, organic waste can form important feedstock resources for the sustainable production of insects (Nyakeri et al. 2017). The distinction between food, food waste and food no longer intended for human consumption is indeed fundamental to define which products can be used as feed material and which product shall be directly classified as organic waste. The European Commission has recently published a document that provides clarifications on the concepts of organic waste and food no longer intended for human consumption in order to valorize the nutrients of food which is, for commercial reasons or due to problems in manufacturing or certain defects, no longer intended for human consumption (European Commission 2018b). According to the European Commission, former food shall not be automatically considered waste but it can be used as feed material provide that specific conditions are met.

Insect farming until the termination of insects shall be considered primary production according to the definition set forth in article 3(17) of Regulation (EC) No 178/2002[14] and as such it is subject to hygiene requirements set forth in Annex 1 to Regulation (EC) No 852/2004 for primary production (EVIRA 2018, 9; FASFC 2018, 5).

In European insect farms, insects are kept in a closed environment, in boxes/cages, where the atmosphere, substrate, water etc. can be controlled. According to

---

[11] According to the Recital (6) of Commission Regulation (EU) 2017/893 of 24 May 2017 amending Annexes I and IV to Regulation (EC) No 999/2001 of the European Parliament and of the Council and Annexes X, XIV and XV to Commission Regulation (EU) No 142/2011 as regards the provisions on processed animal protein "*insects bred for the production of processed animal protein derived from insects are to be considered as farmed animals, and are therefore subject to the feed ban rules laid down in Article 7 and Annex IV to Regulation (EC) No 999/2001 as well as to the rules of animal feeding laid down in Regulation (EC) No 1069/2009*".

[12] Commission Regulation (EU) No 68/2013 of 16 January 2013 on the Catalogue of feed materials, L 29/1 of 30.1.2013.

[13] See Annex III to Regulation (EC) No 767/2009 of the European Parliament and of the Council on the placing on the market and use of feed, L 229/1 of 1.9.2009.

[14] According to article 3(17) of regulation (EC) No 178/2002 "*primary production means the production, rearing or growing of primary products including harvesting, milking and farmed animal production prior to slaughter. It also includes hunting and fishing and the harvesting of wild products*".

the guidelines provided by the authorities, the rearing containers and similar used for the insects have to be manufactured using chemically safe materials. Even though rearing boxes for live insects are not actual packaging and food contact materials, it is considered a good practice to acquire containers for insect rearing that are made from materials suited for food contact (EVIRA 2018, 10). Today, insects living and killing conditions are not regulated at European level: since insects are non-vertebrates, they are out of the scope of the animal welfare directive.[15] Regulation (EC) 1/2005 on the protection of animals during the transport and Regulation (EC) 1099/2009 on the protection of animals at the time of killing do not apply for the same reason. In the future, these provisions would need to be revised to adapt them to insects' mini-livestock.

Insects Processing and Labelling

Insect processing depends on the form in which insects are distributed for human consumption. It is currently possible to find on the market whole insects, whole insects processed into e.g. powder or paste and insects extracts such as protein isolate, fat/oil or chitin (EFSA 2015, 14). Insect manufacturing and processing shall comply with general food law principles as set forth in Regulation (EC) 178/2002 and Regulation (EC) 852/2004 on food hygiene. Regulation (EC) 853/2004 which lays down specific hygiene rules for food of animal origin does not contain any specific provision for foods produced from insects and an update of this regulation will be necessary in the near future to take in consideration the peculiarities of this mini-livestock. Regulation (EC) 2073/2005 on microbiological criteria of foodstuffs, Regulation (EC) 1881/2006 on contaminants and Regulation (EU) 37/2010 on maximum residue levels of pharmacologically active substances will also need to be updated since there are no specific provisions concerning insect products.

Insects and insect-based products shall comply with the general labelling requirements set forth in Regulation (EU) 1169/2011 and other relevant labelling requirement provided in the Union food law.[16] Additional labeling information, such as the description of food, its composition and the condition of use might be required to ensure that consumers are sufficiently informed of the nature and safety of the insect-based product.

According to EFSA, the consumption of insects and insect-based foods may cause allergic reactions and even anaphylactic shock in humans. In particular, many

---

[15] Council Directive 98/58/EC of 20 July 1998 concerning the protection of animals kept for farming purposes, L 221/23 of 8.8.1998.

[16] E.g. According to Commission Implementing Regulation (EU) 2018/775 of 28 May 2018, the Country of origin of the primary ingredient shall be provided when the country of origin or the place of provenance of a food is given, voluntarily or mandatorily, by any means such as statements, pictorial presentation, symbols or terms, and where the country of origin or place of provenance is not the same as that of its primary ingredient.

insect species contain chitin, a naturally occurring polysaccharide of glucosamine, which can be found also in the cell walls of fungi and the exoskeleton of crustaceans (e.g. crabs, lobsters and shrimp). Although chitin and its derivative chitosan (produced industrially via de-acetylation of chitin) are not allergenic individually, it has immune modulatory properties, depending on the administration route and the size of the chitin particles (FAO 2013; Muzzarelli 2010; Lee et al. 2011). In light of this, the use of allergens statements may be required on insect products and allergens precautionary labelling (e.g. may contain) might be advisable on the products processed in the same line used for insect products, when the risk of cross contamination cannot be excluded.

Insects as Food: The US Approach

In the United States, the FDA has devoted significant attention to insects in human food as defects but has given little public attention to insects as human food or as an intentional component of human food (Boyd 2017, 20).

As confirmed by an extensive case law,[17] insects and their parts are mainly considered as pest that may render a food unfit for consumption according to 21 U.S. Code § 342(a)(3) or which presence in food facilities may create a reasonable probability of contamination that renders a food adulterated according to 21 U.S. Code § 342(a)(4). Beside this, the FDA only regulates the use of insects as substances voluntary added to food in 21 U.S. Code § 73.100 which sets forth the product specifications of cochineal: a red color extracted by the dried and ground bodies of cochineal (*Dactylopius coccus costa*) which is largely used by food industry to color a variety of foods.

Despite the lack of a clear regulatory framework, insects and insect-based foods are having a quick diffusion in the US, where it is currently possible to buy different types of proteins bars, snacks and other processed foods. The blooming of this new food business has resulted in a number of inquiries submitted to FDA concerning the regulatory classification of insects as food. The FDA has not taken an official position on this point but its view, as resulting from the "FDA's Standard Response to Entomophagy Inquiries" is that "*Under the Food, Drug and Cosmetic Acts, bugs/ insects are considered food if they are to be used for food or as component of food*" (Ziobro 2015).

The regulatory classification of insects and their part as food implies that insects shall be "*clean and wholesome (i.e. free from filth, pathogens, toxin), must have been produced, packaged, stored and transported under sanitary conditions, and must be properly labeled (Sec. 403)*" (Ziobro 2015). Additionally, insects shall be

---

[17] Among the others United States v. Cassaro, Inc., 443 F.2d 153, 156 (1st Cir. 1971); Golden Grain Macaroni Co. v. United States, 209 F2d 166, 166–68 (9th Cir. 1953). For an extensive analysis of US case law, see Boyd, 2018, 40.

raised specifically for human food following current good manufacturing practices (cGMPs) in order to minimize vector, structural and toxicogenic hazards. Moreover, since insects grown in the wild may be infected with pathogenic micro-organisms or contain harmful metals or pesticides due to human activities (FAO 2013, 119–122), their collection in the wild is not allowed.

The Regulation of Insects as an Intentional Component of Food

The circumstances that insects and insect-based foods are currently available for sale in several American distribution channels, draws attention to their regulatory classification since to be legally placed on the US market, they shall either be approved as food additive or their use shall be generally recognized as safe (GRAS) (LeBeau 2015).

Under 21 U.S. Code § 170.3 CFR any substance the intended use of which results or may reasonably be expected to become, directly or indirectly, a component of food or otherwise affect the characteristics of any food, is a food additive unless its use is Generally Recognized As Safe (GRAS). Additives are subject to premarket approval by the FDA which is not required for substances which use has been generally recognized as safe.

The classification of insects and insect derivatives as a food additive would imply that their use need to be approved by FDA, following the submission of a petition which demonstrates that the substance is safe when added to food. The authorization would consist of a regulation which sets forth the conditions under which the insect species can be safely used. Although the legal recognition of insects as a food additive would have the positive effect of reassuring consumers about their safety, the solution presents several limitations (Boyd 2017). First, to be classified as a food additive, insects cannot be the sole component of a food since "*single component (of a food) does not affect the characteristics of the food in question, rather, it constitutes the food[18]*". Secondly, the additive approval process does not seem suitable to the regulation of insects due the diversity of insect species, each of which would require a separate approval for use. Finally, the approval process is costly and time consuming especially in consideration of the lack of scientific studies and research related to the use of insects as food.[19]

---

[18] United States v. Two Plastic Drums, More or less of an Article of Food, 984 F.2d 814, 818 (7th Cir. 1993).

[19] The lack of scientific research has been pointed out by EFSA which has highlighted the lack of studies concerning the occurrence of human and animal bacterial pathogens in insects processed for food and feed as the lack of information related to the likelihood of human viruses such as norovirus, rotavirus, Hepatitis E and A being passively transferred from feedstock through residual insect gut contents (EFSA 2015, 39).

In summary, the qualification of insects as substances which use has been generally recognized as safe (GRAS) seems to be the most viable solution. There are two alternatives for establishing the GRAS status: experience based on common use in food before January 1, 1958 or a scientific procedure (Fortin, 2017).

For insects, the scientific procedure, based on published and unpublished data, seems to be the most viable solution. The positive aspect of this procedure is that the insects and insect-based foods manufacturer can make their self-determination that the substance is GRAS for the intended use *without having to wait for FDA to review or approve the determination* (Boyd 2017)". Beside the self-assessment performed by the manufacturer, since 1997 the FDA has introduced a voluntary notification procedure where any person can notify the FDA that a particular use of a substance is exempt from the food additive approval requirements based on the notifier's determination that a specific use is GRAS.

The proof of GRAS safety based on scientific procedures is as rigorous for a food additive and requires the same quantity and quality of scientific evidence (Fortin, 2017). It is estimated that the amount of research needed for demonstrate the GRAS status of an ingredient such as the cricket flour would cost around 250 thousand dollars (Lähteenmäki-Uutela et al. 2017).

Since insects have a long history of consumption in many areas of the world, a possible solution is to determine the GRAS status of the substance based on its use prior to January 1, 1958. In this case, the manufacturer making the GRAS determination would have to demonstrate that the substance is generally recognized as safe under the condition of its intended use through the experience based on common use in food. Even if the evidence is based upon generally available data and information and it is less-resource intensive than the ones required for additives and GRAS substances based on scientific procedures, the FDA seems to be "reluctant" to rely on the common use of substance outside the United States (Hutt et al. 2014).

When placed on the market, insect products shall be properly labeled indicating both their common and scientific name to avoid misleading consumers on the nature of the food they are consuming.

Conclusions

The analysis of the regulatory framework both in the European Union and US shows that insects and insect products have attracted the attention of the legislator that has only recently clarified their legal status and the rules for their placing on the market. This is only the first step since the food legislation needs to be updated to take into consideration the peculiarities of this mini-livestock and the new risks that it may pose.

References

Bodenheimer FS (1951) Insects as human food. A chapter of the ecology of man. Junk W, editor, The Hague

Boyd MC (2017) Cricket soup: a critical examination of the regulation of insects as food. Yale Law and Policy Rev 36(1):17–81

Dobermann D et al (2017) Opportunities and hurdles of edible insects for food and feed. Nutrition Bulletin 42:293–230

EFSA (2015) Scientific Opinion on a risk profile related to production and consumption of insects as food and feed. EFSA Journal 13(10):4257., 60 pp. https://doi.org/10.2903/j.efsa.2015.425

EFSA (2016a) Guidance on the preparation and presentation of an application for authorization of a novel food in the context of Regulation (EU) 2015/2283. EFSA J 14(11):4594. 24 pp

EFSA (2016b) Guidance on the preparation and presentation of the notification and application for authorisation of traditional foods from third countries in the context of Regulation (EU) 2015/2283. EFSA J 14(11):4590. 16 pp

EFSA (2018) Administrative guidance on the submission of applications for authorisation of a novel food pursuant to Article 10 of Regulation (EU) 2015/2283, EFSA supporting publication 2018:EN-1381. 22 pp. https://doi.org/10.2903/sp.efsa.2018.EN-1381

European Commission (2018a) Summary of ongoing applications and notifications, available at the following web address: https://ec.europa.eu/food/safety/novel_food/authorisations/summary-ongoing-applications-and-notifications_en. Last accessed on 21 Dec 2018

European Commission (2018b) Guidelines for the feed use of food no longer intended for human consumption, 2018/C 133/02

Evira (Finnish Food Safety Authority) (2018) Insects as food (10588/2). https://www.evira.fi/globalassets/tietoa-evirasta/lomakkeet-jaohjeet2/elintarvikkeet/eviran_ohje_10588_2_uk.pdf. Last accessed on 21 Dec 2018

FAO (2013) Edible insects: future prospects for food and feed security, Food and Agric. Org. United Nations, available at the following web address http://www.fao.org/3/i3253e/i3253e.pdf. Last accessed on 24 June 2019

Federal Agency for the Safety of the Food Chain (FASFC) (2014) Food safety aspects of insects intended for human consumption (Sci Com dossier 2014/04; SHC dossier n° 9160), available on the web site: https://www.afsca.be/scientificcommittee/opinions/2014/_documents/Advice14-2014_ENG_DOSSIER2014-04.pdf

Federal Agency for the Safety of the Food Chain (FASFC) (2016) Circulaire relative à l'élevage et à la commercialisation d'insectes et de denrées à base d'insectes pour la consommation humaine, available on the web site: http://www.afsca.be/denreesalimentaires/circulaires/_documents/2016-04-26_circ-ob_FR_insectes_V2_clean.pdf

Federal Agency for the Safety of the Food Chain (FASFC) (2018) Circulaire relative à l'élevage et à la commercialisation d'insectes et de denrées à base d'insectes pour la consommation humaine, available on the web site: http://www.favv-afsca.be/denreesalimentaires/circulaires/_documents/2018-11-05_omzendbriefinsectenv3FR_clean.pdf. Last accessed on 21 Dec 2018

Finke MD et al (2015) The European Food Safety Authority scientific opinion on a risk profile related to production and consumption of insects as food and feed. J Insects Food Feed 1:245–247

Fortin N (2017) Food regulation: law, science, policy, and practice, 2nd edn. Wiley, New Jersey

Hutt et al. (2014) Food and drug law: cases and materials, University casebook series, 4th edn. Foundation Press, Sunderland

Jongema (2017) List of edible insects of the world (April 1, 2017) – WUR. Available at: http://www.wur.nl/en/Expertise-Services/Chairgroups/Plant-Sciences/Laboratory-of-Entomology/Edible-insects/Worldwide-species-list.htm. Last accessed on 27 Dec 2018

Lähteenmäki-Uutela A et al (2017) Insects as food and feed: laws of the European Union, United States, Canada, Mexico, Australia and China. European Food Feed Law Rev 1:22–36

LeBeau A (2015) Insect protein: what are the food safety and regulatory challenges?, available on the website: http://burdockgroup.com/insect-protein-what-are-the-food-safety-and-regulatory-challenges/. Last accessed on 25 June 2019

Lee CG, Da Silva CA, Dela Cruz CS, Ahangari F, Ma B, Kang M-J, He C-H, Takyar S, Elias JA (2011) Role of chitin and chitinase/chitinase-like proteins in inflammation, tissue remodeling, and injury. Annu Rev Physiol 73(1):479–501

Livsmedelsverke (2017) Inget kryphål i lagen för insekter som mat i Sverige, available at the following web address: https://www.livsmedelsverket.se/om-oss/press/nyheter/pressmeddelanden/inget-kryphal-i-lagen-for-insekter-som-mat-i-sverige. Last accessed on 24 Jun 2019

Ministero della Salute (2013) Controlli ufficiali in merito all'uso di insetti in campo alimentare con specifico riferimento all'applicabilità del reg. (CE) 258/97 sui "novel food", 29.10.2013

Muzzarelli R (2010) Chitins and chitosans as immunoadjuvants and non-allergenic drug carriers. Mar Drugs 8(2):292–312

Nyakeri EM et al (2017) Valorization of organic waste material: growth performance of wild black soldier fly larvae (Hermetia illucens) reared on different organic wastes. J Insects Food Feed 3:193–202

Office for Risk Assessment & Research (ORAR) (2014) Advisory report on the risks associated with the consumption of mass reared insects, available on the website: https://www.research-gate.net/publication/277716517_Advisory_report_on_the_risks_associated_with_the_consumption_of_mass_reared_insects. Last accessed on 25 Jun 2019

Paganizza V (2016) Eating insects: crunching legal clues on entomophagy. Rivista di Diritto Alimentare 1:16–14

Pisanello D et al (2018) Novel food in the European Union. Springer International Publishing, Switzerland

Testa M et al (2016) Ugly but tasty: a systematic review of possible human and animal health risks related to entomophagy. Crit Rev Food Sci Nutr 57(17):3747–3759

Ziobro (2015) FDA Ctr. for Food Safety & Applied Nutrition, Presentation at the Institute of Food Technologists (IFT), Regulatory Issues, Concerns and Status of Insect Based Foods and Ingredients, Chicago, July 13, 2015

Index

A
Acceptance, viii, 7, 13–22, 28–42, 46–53, 58–61, 63–65, 74–84
Alternative proteins, vii, 20–22, 28, 46, 51
Animal origin food, 113
Aquaculture, 74, 78, 83

B
Behaviour, 29, 31, 37, 39, 40, 46, 59, 83

C
Consumer behavior, 30
Crustaceans, 88–91, 99, 100, 114

D
Disgust, viii, 18–22, 28, 39–41, 47–49, 51–53, 59–62, 67

E
Edible insects, vii–ix, 2, 15–22, 28, 46, 88, 106
Eggs, 2, 81
Emotions, viii, 39, 41, 47, 49, 58, 67
Entomophagy, vii, viii, 6–7, 12–22, 29, 33, 39, 42, 46, 58–62, 67, 88–100, 106, 108, 114
Europe, vii, viii, 12–22, 33, 37, 41, 42, 66, 83
European food safety authority (EFSA), viii, 4–6, 19–22, 109–111, 113

F
Feed, vii, 16–22, 46, 74, 109
Food, vii, viii, 1, 12–22, 28, 46, 82–83, 88, 106
Food allergy, 88, 90–92, 100
Food history, 15–22

G
Generally recognized as safe (GRAS) status, 115, 116

H
House dust mites, 88–91, 99, 100

I
Insect-based foods, 29, 31, 37, 38, 41, 61, 66, 67, 113–116
Insects, 1, 12, 28, 46, 74, 88, 106

M
Meat, vii, 28, 34, 36, 37, 46, 47, 50, 51, 53, 62, 64, 74, 78, 81, 108

N
Neophobia, 39, 47, 49, 59, 67
Novel, vii, viii, 5–6, 17–22, 28, 29, 36, 41, 46, 47, 83, 106–111
Novel foods, vii, viii, 3, 28, 46, 82, 88, 106

© Springer Nature Switzerland AG 2019
G. Sogari et al. (eds.), *Edible Insects in the Food Sector*,
https://doi.org/10.1007/978-3-030-22522-3

Printed in the United States
By Bookmasters